METHUEN'S MONOGRAPHS
ON APPLIED PROBABILITY AND STATISTICS

General Editor: M. S. BARTLETT, F.R.S.

MODELS IN REGRESSION
and Related Topics

Models in Regression

and Related Topics

PETER SPRENT

Senior Lecturer in Statistics,
Department of Mathematics,
University of Dundee

METHUEN & CO LTD
11 NEW FETTER LANE, LONDON EC4

First published 1969
© *Peter Sprent* 1969
Printed in Great Britain by
Willmer Brothers Limited
Birkenhead

SBN 416 14830 1

Distributed in the U.S.A.
by Barnes and Noble, Inc.

Contents

CONTENTS

References 181

Preface

This book gives a more detailed treatment of regression and related topics in curve and surface fitting in the presence of random variation than that given in most general statistical tests.

The treatment is, however, by no means exhaustive. The aim is to provide a broad survey for anyone who wants to study these topics in depth, but who is deterred by the vastness of the literature and the mass of notational differences and subtleties of argument that are soon encountered.

Two broad classes of reader are envisaged. The first is the student of mathematical statistics with special interests in this field. The second is the research worker who feels these methods may be useful, but who wants guidance in the choice of appropriate models for his particular problem.

Although detailed mathematics has been kept to a minimum, an undergraduate knowledge of calculus is assumed and matrix notation is used where appropriate. It has been assumed that the reader has a basic statistical knowledge of the type that would be acquired in an introductory course of some fifty lectures, or by the reading of a standard introductory text.

As one of the aims of the book is to show how models are related to one another, as well as the differences between them, the divisions between topics are not clear-cut. However, the first four chapters deal with the straight line entirely, while linear functions of many variables, and curves other than straight lines, are dealt with in chapters 5 and 6. These earlier chapters deal with basic problems of selection of models and fitting them to data. Applications of the methods are dealt with in chapters 7 and 8. Chapter 9 deals with several miscellaneous topics, and chapter 10 provides a very brief introduction to non-linear models.

Computational details have generally been omitted, but references are given to the many sources of information on this aspect.

ix

The selection of topics for inclusion reflects to some extent my own interests, although I have tried to concentrate on models likely to be of fairly wide applicability, and to indicate where important topics have not been fully developed.

Exercises, some worded in very general terms, have been included at the end of all but the first chapter. These vary from almost trivial additions to matters dealt with in the book, to suggestions for developing ideas that are only briefly mentioned in the text. Where appropriate, references have been given, and it is hoped the interested reader will follow these up.

I am grateful to the Director of East Malling Research Station, Dr F. R. Tubbs, C.B.E., and to Professor H. A. David, Editor of *Biometrics*, for permission to use the data in tables 8.1 and 8.3. I would also like to thank several fellow statisticians for stimulating discussions on the subject matter of this book, and in particular Professor D. R. Cox and Mrs Joyce Snell of Imperial College, and Dr S. C. Pearce of East Malling Research Station for helpful and encouraging comments on an early draft. The editor of this series was also kind enough to make a number of constructive suggestions, and his help is gratefully acknowledged. The responsibility for any shortcomings that remain rests squarely upon my shoulders.

PETER SPRENT

University of Dundee
January 1969

CHAPTER 1

Introduction

1.1. The problem

Most people associate regression techniques with the fitting of straight lines, curves, or surfaces, to sets of observations, where the fit is for one reason or another imperfect. Nearly all standard statistical textbooks include at least one chapter on regression, and most recent issues of the major statistical journals contain one or more papers on the topic.

Curves or surfaces may be fitted for such obvious purposes as summarizing trends in data, or for purposes of prediction, or to establish what have sometimes been termed natural laws (or lawlike relationships); there are one or two less obvious applications, and one of these we shall meet in later chapters is the control problem.

This book is concerned with the mathematical models appropriate to the various situations where regression and related techniques are used. The aim will be to illustrate both theoretical and practical differences between various models in common use. Little or no guidance is given about computation as there are already several texts that cover this aspect very adequately.

The straight line is the simplest curve that can be fitted to a set of n paired observations $(x_1, y_1), (x_2, y_2) \ldots (x_n, y_n)$. A problem of fitting occurs only if the fit is for some reason imperfect. To be a statistical problem there must be some random element present in the data which leads to this inexactitude of fit. It is the nature of this random element that determines the appropriate method of fitting, i.e. of estimating the *constants* or *parameters* in the equation.

In passing we note that the statistician uses the term parameter in a rather different sense to the mathematician when the latter speaks of parametric specification. When the statistician estimates, from a sample of values of x and y, the constants α, β in the equation

$$y = \alpha + \beta x$$

he thinks of α, β as the parameters. The mathematician thinks of ρ as a parameter when he specifies a straight line in the parametric form

$$x = x_1 + l\rho, \qquad\qquad y = y_1 + m\rho,$$

where l, m are the direction cosines of the line and it passes through the point (x_1, y_1). Throughout this book we shall use the word parameter in the statistical sense.

1.2. Scope and treatment level

The subject will be treated in a way that assumes the reader has a working knowledge of statistical techniques at or above the level implied by Wetherill (1967) or a similar text. An elementary knowledge of calculus is assumed, and matrix notations are used. Readers unfamiliar with matrix algebra will find a useful summary in the appendix to Anderson (1958), and a more detailed treatment is given by Searle (1966).

As this book is largely concerned with a formulation and discussion of various models and the practical implications of differences between them, mathematical detail that is readily available elsewhere has often been omitted; it has been included when it is not easily available elsewhere, or when it illustrates an important characteristic of a model or techniques for handling it.

A number of different mathematical models will be discussed, and some of these find applications in more than one field such as the physical or biological sciences, technology, economics, psychology and education. Many of the examples quoted in this book arise in biology, as it is in this field that the author has had most experience.

It is hoped that the user will find the volume a useful guide when formulating a model, although sometimes his problem may require a model that is not already formulated here or elsewhere in the literature.

1.3. Notation and terminology

Standard statistical texts usually deal with rather special models when discussing straight line regression. These are usually called *bivariate* regression and *least squares* regression. As many of the computational techniques in applying the theory are identical for these two models there is often confusion about the logical distinction between them; moreover, many students complete their formal statistical training in the belief that these are the only models of importance in this field. The notation and terminology are often confusing.

Some would claim that all models considered in this book are regression models, and that all the problems considered are regression problems. Others would claim that we are dealing with several logically distinct concepts called regression, functional relationship and structural relationship. The author's own view is that there is a continuum of models and techniques having the common feature that they give rise to relationships between random variates and/or mathematical variables that are expressible in the form of equations.

One or two terms and symbols are used more or less universally in this field, but apart from these nearly every major writer on the subject uses different notations, the confusion being added to by different writers using the same symbols with different meanings. Extreme care must therefore be taken to see that each author's notation is understood. No claim of superiority is made for that used in this book; it has been adopted after considerable thought because it seems to be workable and fits in reasonably well with other widely used notations.

Some symbols, notably x_i, y_i are very overworked in regression studies; x_i, y_i nearly always denote observed values, *but* whereas both are sometimes random variates, there are many cases where one of these is a mathematical variable. With a little care confusion can be avoided; altering this convention would have complicated the mathematics and introduced unconventional notation.

It will be necessary to distinguish between RANDOM VARIATES and MATHEMATICAL VARIABLES. For simplicity we shall refer to the former simply as VARIATES and reserve the term VARIABLE for a mathematical variable, an entity with no associated probability distribution. We shall occasionally retain the adjectives 'random' and 'mathematical' for emphasis. In reading other texts, note that this convention is by no means standard.

In practice we are nearly always concerned with deducing something about a model using observational data that may consist entirely of random variates, or of a mixture of random variates and mathematical variables. A variate itself may sometimes be composed of a variable plus a random component, the latter often, but not necessarily, having a connotation of 'error'.

We sometimes observe fixed or deliberately chosen values of what is basically a random variate. These non-randomly chosen values are best regarded in regression theory as values of a mathematical or observational variable, as the 'random' element is destroyed by the selection process. For example, the heights of all army recruits are approximately normally distributed, but if we are instructed to select 10 recruits, of which 2 are to be of height 170 cm, 4 are to be of height 174 cm, 3 are to be of height 177 cm, and one is to be of height 180 cm, simply by going through the records until we have the requisite numbers of recruits of each height, this selected group in itself gives no information on the distribution of the random variate height; the values 170, 170, 174, 174, 174, 174, 177, 177, 177, 180 are best regarded as variable values (as distinct from variate values). If the 10 recruits had been selected *at random*, and their heights x_i recorded, then the x_i would have been realized values of a variate.

1.4. Galton's concept of regression

The term *regression* was introduced to Statistics by Francis Galton in a series of papers, the most famous being Galton (1886), to describe an hereditary phenomenon. He noted that although there was a tendency for tall parents to have tall children, and for short parents to have short children, the distribution of heights in a population did not change from generation to generation. In a very detailed paper he showed that this constancy could be explained by a tendency for the mean height of children who had parents of a given height to *regress* or move towards the population average height.

Galton's paper is well worth reading as an example of the many practical considerations that have to be kept in mind in collecting and interpreting data. Not only does he discuss possible inaccuracies in measurement, but he considers, for instance, the fact that the heights of both parents will influence the height of a child, and is thus led to use what he calls mid-parent height. Surprisingly, this is not the mean height of the parents, for his studies showed the influences of the two parents to be unequal, and before averaging he multiplied each female height by 1·08.

4

Galton found that the mean heights of children for each of a set of given mid-parent heights lay approximately on a straight line. To him, the important point that justified his calling this a *regression* line was that the slope was less than unity (implying the regression, or movement towards the population mean, of the heights of children born to parents with a given mid-parent height). Had the slope been greater than or equal to unity the means would not have 'regressed'.

1.5. The modern concept of regression

Modern statistical usage of the term regression places no such restriction on the slope, and the term is associated with curves other than straight lines. Some writers still use the word regression only in situations where a mean is the concept of interest. We shall see that methods of computation can often be the same for logically distinct situations whether or not a mean is the prime concept of interest, so there seems little point in a distinction that calls one a regression problem and the other something else. It is the author's opinion that in a particular situation the important thing is to appreciate the form of the model being used, and to assess its relevance to the problem under consideration. This implies, among other things, that when phrases such as 'using standard regression techniques' or 'this reduces to a regression problem' occur in the literature, the reader must satisfy himself as to how that author defines regression.

In this book the view is taken that any problem involving the expression of the mean value of a variate as a function of other variates or variables is certainly a regression problem. Other terms such as *functional* and *structural* relationship will be used when considering certain law-like relationships. These will involve examination of problems that other writers have referred to as 'regression with both variables subject to error'. We shall meet problems that do not fit neatly into one of the above classes, and it is for this reason that we take the view already expressed that models form a continuum rather than a small number of distinct classes.

Our random variates will frequently be assumed to have a normal distribution. The possibility of other distributions arising in practice cannot be entirely ignored. The reader is urged to consider in the various cases discussed, to what extent results depend upon the assumption of normality; for instance, assessment of the accuracy of a regression line is based essentially upon the concept that deviations from the line are normally distributed.

The methods of estimation we shall use will nearly always be either maximum likelihood or least squares, generalized when appropriate. The two are often equivalent, especially where normality is involved. Brief mention will be made of other methods of estimation. These are sometimes less efficient, but may also be less laborious; but in the computer era, difficulties of calculation are less serious than they were even a decade ago. The real danger with computers is that too many useless calculations may be performed. This is one reason why an understanding of mathematical models and their *relevance* to a particular objective is important.

CHAPTER 2

Bivariate Normal and Least Squares Regression

2.1. The two cases of classical regression

In this chapter we discuss the two cases of regression found in most standard statistical texts. We deal fairly briefly with the estimation problems associated with these models. As a clear understanding of these cases is helpful in studying later chapters, it is hoped that readers who are already familiar with much of the content of this chapter will at least skim through it with the aim of ensuring that they understand what assumptions are being made and their implications.

Technical details about application of these models has been kept to a minimum, and for this the reader is advised to consult texts such as Williams (1959), Acton (1959), Draper and Smith (1966) or the relevant chapters of Kendall and Stuart (1967). The more theoretical aspects of regression are discussed by Plackett (1960).

2.2. Bivariate normal regression

The first case met in many standard texts is bivariate regression. If X, Y are variates with joint probability density function $f(x, y)$, the mean value of Y conditional upon X taking the value x is defined as

$$E(Y|X = x) = \frac{\int\limits_{-\infty}^{\infty} y f(x, y) \, \mathrm{d}y}{\int\limits_{-\infty}^{\infty} f(x, y) \, \mathrm{d}y}. \qquad (2.1)$$

This is a function of x, and by varying x a curve is obtained which is known as the regression curve of the variate Y upon the variate X.

If X, Y are normally distributed with means μ_X, μ_Y respectively, standard deviations σ_X, σ_Y and correlation coefficient ρ it is well known (see, e.g. Kendall and Stuart (1967), or Anderson (1958)), that the conditional distribution of Y given $X = x$ is normal with mean

$$E(Y|X = x) = \mu_Y + \frac{\rho\sigma_Y}{\sigma_X}(x - \mu_X)$$

and variance $\sigma_Y^2(1-\rho^2)$. If x is allowed to vary, the line with equation

$$y = \mu_Y + \frac{\rho\sigma_Y}{\sigma_X}(x - \mu_X) \tag{2.2}$$

is the equation to the locus of the conditional mean and is clearly a straight line passing through (μ_X, μ_Y) with slope $\rho\sigma_Y/\sigma_X$ and this is the regression of Y upon X. Note that y in (2.2) is not a realized value of the (random) variate Y, but is a (mathematical) variable which, for given x, takes the value $E(Y|X = x)$. The regression of X on Y is defined as the locus of $E(X|Y = y)$, and for a bivariate normal distribution it has the equation

$$x = \mu_X + \frac{\rho\sigma_X}{\sigma_Y}(y - \mu_Y). \tag{2.3}$$

Like (2.2), this line passes through (μ_X, μ_Y) but it has a different slope. It is convenient to write the slopes of (2.2) and (2.3) as

$$\beta_1 = \frac{\rho\sigma_Y}{\sigma_X} = \frac{\text{Cov}(X, Y)}{\sigma_X^2}$$

and

$$\beta_2 = \frac{\sigma_Y}{\rho\sigma_X} = \frac{\sigma_Y^2}{\text{Cov}(X, Y)}.$$

Note that $\beta_1 = \beta_2$ if and only if $\text{Cov}^2(X, Y) = \sigma_X^2\sigma_Y^2$, i.e. $\rho^2 = 1$. In this case the bivariate normal distribution is said to be singular, and it is concentrated entirely on a straight line of slope $\beta = \beta_1 = \beta_2$.

The above normal distribution regression theory is straight-forward. Inference problems arise when the parameters in equations (2.2) or (2.3) have to be estimated from random samples of observations from a bivariate normal population. Using maximum likelihood, or minimum variance unbiased estimators, for μ_X, μ_Y, $\text{Var}(X)$, $\text{Var}(Y)$ and $\text{Cov}(X, Y)$ based on a sample of n observations (x_i, y_i) we find estimators $\hat{\mu}_X$, $\hat{\mu}_{,Y}$ $\hat{\beta}_1$, $\hat{\beta}_2$ for μ_X, μ_Y, β_1, β_2, namely

$$\hat{\mu}_X = n^{-1}\sum x_i, \quad \hat{\mu}_Y = n^{-1}\sum y_i, \hat{\beta}_1 = \frac{\sum x_i y_i - (\sum x_i)(\sum y_i)/n}{\sum x_i{}^2 - (\sum x_i)^2/n},$$

$$\hat{\beta}_2 = \frac{\sum y_i{}^2 - (\sum y_i)^2/n}{\sum x_i y_i - (\sum x_i)(\sum y_i)/n}.$$

Note that we may write the estimated regression equation (2.2) in the form

$$y = \hat{\alpha}_1 + \hat{\beta}_1 x$$

where

$$\hat{\alpha}_1 = \hat{\mu}_Y - \hat{\beta}_1 \hat{\mu}_X,$$

with an analogous form for (2.3).

2.3. An example of bivariate normal regression

Figure 2.1, based upon data similar to that collected by Galton in the study discussed in section 1.4, shows a scatter diagram of some typical mid-parent heights (x) and eldest son's heights (y). It is well established that in large populations each set of heights approximately follows the normal distribution; thus if n sets of observations (x_i, y_i), $i = 1, 2, \ldots, n$, are taken at random they represent a sample of n paired values from a bivariate normal population. In section 2.2 we gave appropriate formulae for estimating the parameters in the regression equation for Y upon X from such a set of sample values.

Regression here tells us something about the *mean* height of sons who have a given mid-parent height. The line in figure 2.1 represents the regression line estimated from the plotted data. It enables us to estimate the average height of an eldest son in families with any given mid-parent height. For an individual family it will not, of course, tell us (or predict) the exact height of an eldest son. It hardly needs a knowledge of genetics or biology to tell us that this is in conformity with experience, because the height of an individual is determined by many genetical factors, as well as environmental conditions such as climate, housing, nutrition and the effects of disease and accidents.

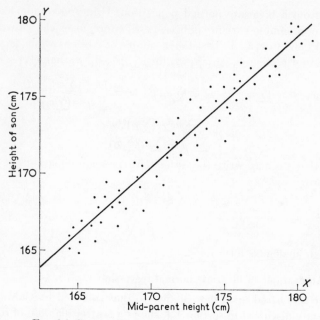

FIG. 2.1. Scatter diagram of son and mid-parent heights

The regression line summarizes a tendency, and it would obviously be useful if, having got the line, we could go on to make statements of the form:

'If we consider all parents with mid-parent height x cm, then numbers δ_1, δ_2 (not necessarily different from one another) can be found such that 95 per cent of their eldest sons will have heights between $y - \delta_1$, and $y + \delta_2$ cm, where y, δ_1 and δ_2 can be determined.'

The reader will recognize statements of this type to be of common occurrence in statistical inference, and there are indeed techniques that enable us to make this sort of statement in regression problems.

When we estimate a regression line from data, what we get is an *estimate* of the true population regression line. If the estimated slope is $\hat{\beta}$ we can write down, say, 95 per cent confidence limits for β in the form $\hat{\beta} \pm \delta$, where δ can be determined.

Some caution is needed in specifying the population for which inferences are valid. If the study were made on a group of English families, the results clearly may not apply to Chinese families. Statistical inferences about populations only apply strictly to the population from which the data are a random sample. We use the word population in a statistical sense.

We note that the regression of height of eldest son on mid-parent height represents a clear cause-effect relationship between heights of parents and heights of children. Genetic factors that determine whether parents are or are not tall are passed on to the children, causing them to be tall if the parents are. There is a lack of symmetry in the sense that while heights of parents influence height of children, a child's height does not affect the height of its parents!

For the data in figure 2.1, computationally there is nothing to stop us from calculating an estimated regression line of parents' heights, X, on children's heights, Y. It has the practical use of telling us something about the mean heights of parents who have sons of a *given* height, but does not carry any implication that a child's height causes the parents' heights. Anscombe (1967) has criticised modern usage of the word regression, but remarks that he supposes 'it is eighty years too late to complain.' The use of regression in senses vastly different to that envisaged by Galton is now firmly established.

Suppose that, after change of scale if necessary, the data plotted in figure 2.1 represent not heights of humans, but lengths of elephants' legs, x, and elephants' trunks, y. The observed values may again be regarded as realized values of random variates X, Y. We may compute the regression of Y on X, or that of X on Y. From the biological point of view the situation differs from that for the parent – son relationship, since we can no longer say there is a causal relationship between trunk length and leg length. The tendency for both to be long or both to be short is a result of genetical, and perhaps also environmental factors, influencing the growth of both in a similar way; neither 'causes' the other, but each measurement has the value it does as a result of the influence of common factors. In the situation discussed here either regression equation may be useful for purposes of prediction, a use of regression equations that is taken up more fully in section 2.6.

In the light of the remarks on notation in section 1.3 it is worth noting that in this chapter we have already used y with three different meanings, (*i*) as a dummy variable of integration in (2.1),

11

(*ii*) as a mathematical shorthand for $E(Y|X = x)$ in (2.2) and (*iii*) with appropriate suffices as realized values of a random variate Y. The reader is again urged to consider the precise meaning of symbols *each* time they are used.

2.4. Regression on a mathematical variable

The model specified is one in which, for a given value of a (mathematical) variable x, say x_i, the value y_i of a (random) variate Y is given by

$$y_i = \alpha + \beta x_i + \varepsilon_i \qquad (2.4)$$

where ε_i is a variate with zero mean. Note that it is convenient when using italic letters of the Roman alphabet to refer to variates or variables to reserve the capital for the former and the lower-case letter for the latter. There would be a lot to be said for extending this distinction to realized values. However, the use of (x_i, y_i) for observed values is so common that we have adhered to it in this book, specifying the nature – variate or variable – in each specific problem.

If we assume the ε_i in (2.4) to be $N(0, \sigma^2)$ and further assume ε_i to be independent of ε_j if $i \neq j$, then the observations can be used to estimate α and β by the *method of least squares*. This procedure is due to Gauss (1809), although there are indications that the method may have been used earlier. An interesting historical account of developments in this field has been given by Seal (1967). Note that the assumptions about (2.4) imply that for any given x there exists a random variate Y that is $N(\alpha + \beta x, \sigma^2)$. With our assumptions, least squares is easily shown to be equivalent to maximum likelihood. The latter method of estimation is discussed in many statistical texts. See, e.g. Hoel (1962) or Mood and Graybill (1963).

The likelihood function for α, β, σ^2 based upon a sample of n paired observations (x_i, y_i) according with the above model is

$$L = \frac{1}{(2\pi\sigma^2)^{n/2}} \exp\left\{\frac{-\sum(y_i - \alpha - \beta x_i)^2}{2\sigma^2}\right\}.$$

Taking logarithms to the base e,

$$L^* = \ln L = -\frac{n}{2}\ln(2\pi) - \frac{n}{2}\ln \sigma^2 - \frac{\sum(y_i - \alpha - \beta x_i)^2}{2\sigma^2}.$$

L will be a maximum for the same values of α, β, σ^2 that make L^* a maximum. Differentiating L^* with respect to α, β, σ^2, setting these derivatives to zero, and solving the resultant equations, gives the maximum likelihood estimators of the parameters, namely

$$\hat{\alpha} = (\textstyle\sum y_i)/n - \hat{\beta}(\textstyle\sum x_i)/n, \tag{2.5}$$

$$\hat{\beta} = \frac{\sum x_i y_i - \sum x_i \sum y_i/n}{\sum x_i^2 - (\sum x_i)^2/n}, \tag{2.6}$$

$$\hat{\sigma}^2 = \frac{\sum(y_i - \hat{\alpha} - \hat{\beta} x_i)^2}{n}$$

$$= \frac{1}{n}\{\textstyle\sum y_i^2 - (\sum y_i)^2/n - \hat{\beta}(\sum x_i y_i - \sum x_i \sum y_i/n)\}. \tag{2.7}$$

These are well known results and (2.5) and (2.6) are of exactly the same form as the corresponding estimators in bivariate normal regression. In exercise 2.1 the equivalent of $\hat{\sigma}^2$ for bivariate regression is discussed.

As with maximum likelihood estimators of variance in some other situations, $\hat{\sigma}^2$ turns out to be a biased estimator of σ^2, but a small adjustment leads to the unbiased estimator

$$\hat{\sigma}^{*2} = n\hat{\sigma}^2/(n-2).$$

In future we shall use $\hat{\sigma}^{*2}$ as just defined instead of $\hat{\sigma}^2$, but for typographical convenience we shall drop the asterisk. Effectively this means that in future

$$\hat{\sigma}^2 = \frac{\sum(y_i - \hat{\alpha} - \hat{\beta} x_i)^2}{n-2}. \tag{2.8}$$

A commonly used notation is to write $\hat{\alpha} = a$, $\hat{\beta} = b$ and $\hat{\sigma}^2 = s^2$. Note that in (2.7) $\sum y_i^2 - (\sum y_i)^2/n$ is the total sum of squares about the mean for the observed values of Y and has $n-1$ degrees of freedom, and $\hat{\beta}\{\sum x_i y_i - (\sum x_i \sum y_i)/n\}$ is the sum of squares accounted for by fitting $\hat{\beta}$ and this has one degree of freedom. Note that $\hat{\alpha}$ and

$\hat{\beta}$ do not depend upon σ^2, and the same estimators of α, β would have been obtained if we had minimized

$$E = \sum(y_i - \alpha - \beta x_i)^2. \tag{2.9}$$

Now E is the sum of squares of deviations measured in the direction of the y-axis of individual observations from the line $y = \alpha + \beta x$. The choice of α, β in (2.9) so as to minimize E is the celebrated *Principle of Least Squares*. Note that for the model associated with (2.4) where x is a variable rather than a variate, there is no meaning attachable to the regression of x on Y, analogous to the regression of X on Y in the bivariate case.

2.5. Examples of regression on a mathematical variable

Figure 2.2 illustrates a problem where the model associated with (2.4) is appropriate. Here x represents the age in years of boys, and y their heights. Clearly, the heights of boys have been measured at fixed ages, each boy having his height measured on his birthday.

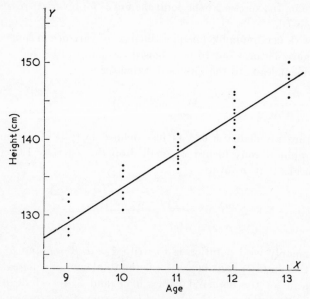

Fig. 2.2. Heights of children of given age measured on their birthdays

We suppose that each measurement is made on a different boy. Here age has the role of an observational variable rather than a variate, as we have selected certain values of it, and measured the heights of boys having these specific values of x. If the boys measured at *each* age can reasonably be regarded as a random selection of boys *of that age* from some population, and height is normally distributed in that population, then the observed y corresponding to any given x will be normally distributed about the mean height for that age. The straight line in figure 2.2, fitted by least squares, gives an indication of how, on average, heights of boys change with age.

The practical man assumes, largely on the basis of experience, that if he now measures the heights of boys aged 10·2 or 11·7 or 8·9 years, that he would get a scatter of points centred about this line for each appropriate x value. He would indeed be surprised if he measured the height of, say, 10 children selected at random from those aged 10·2 years and found them all to be under 90 cm. He also knows it would be meaningless to extrapolate, and use his straight line to estimate the mean height of men aged 40 years! This is very obvious in this example because experience tells us that the straight-line relationship between age and height of children does not extend into the adult age range; but it is surprising how many experimenters forget limitations on the range of validity of mathematical models in more complicated situations. We also point out that a line such as that in figure 2.2, obtained for normal children, cannot be used to make inferences about the heights of children in a school specializing in the teaching of dwarfs. Further, if the relationship were known to hold well for English children, there is no guarantee that it will be satisfactory for French, Canadian, Mexican or Chinese children.

If, instead of measuring the height of children at fixed and predetermined x values, the children had been selected *at random* from all children in the population, the scatter of points would have looked more like that in figure 2.1 (with appropriate choices of scale for x and y). The sample of ages in this case could hardly be regarded as a sample from a normal distribution, for over the school age period our population may show something approaching a uniform distribution of ages. In practice it is still convenient to regard age as a (mathematical) variable in these circumstances in view of the similarity in computational methods for the two regression situations considered in sections 2.2 and 2.4. The important practical difference between the models occurs when we want to estimate the value of x likely to have given rise to an observed y. This problem is discussed in section 7.2.

When the models give rise to the same computing procedures and conclusions it is often more convenient to select children of *fixed* ages for determination of relationships such as that between age and height or age and weight. For example, to spread the work throughout the year a local health authority may conduct a medical check on children on their birthday each year, and this would provide records of their height if it were measured as part of the check.

There is a tendency in the literature to speak of the ε_i in (2.4) as error. In an example such as the age-height relationship this implies that, if we ignore 'errors of measurement', there is some imperfection in an individual if his height departs from the average for his age. Biologists in general, and more particularly champions of individuality, would rightly deplore such an attitude. Variability and individuality are fundamental characteristics of living organisms, and the ε_i really represent departures from the straight line implicit in the model. In this book when dealing with models containing random elements analogous to the ε_i in (2.4) we shall usually refer to these as *departures* rather than errors. There are, of course, situations where the ε_i are indeed errors in the sense of imperfections in measurement or recording.

Many situations arise in the experimental sciences where one hopes the ε_i will always be small, preferably with mean zero. In this situation we hope that the relationship $y = \alpha + \beta x$ is very nearly an exact relationship. A typical case is presented by a chemist who wishes to calibrate a new process for quick determination of the amount of some constituent in a material with a view to replacing a laborious or expensive method by the new process.

Ideally, if x is the true amount of a constituent present he would like the new method to give an estimate y such that $y = x$. In practice this seldom happens. If we assume the old procedure gives the true value x (this being an observable variable) then y will almost certainly have some sort of 'error' which the chemist will want to estimate and allow for as far as possible. He will certainly want to 'correct' for any systematic differences between y and x, and to know something about the magnitude of 'random' differences.

The chemist may find that $y = x + \delta$. If he is very lucky he may find that δ is always a constant, irrespective of the value of x. There is then no statistical problem as once any two of x, y, δ are known the other can be written down. He is much more likely to find that δ is a (random) variate, and if he is lucky he may find that it is normally distributed about a zero mean. Then his observation y, corresponding to a given x, would be a value of a random variate Y that is $N(x, \sigma^2)$. This would be a situation where δ could truly be regarded as error, for the chemist's goal is clearly to make $\delta = 0$. There will be situations where δ is composed of a constant with an added random error. It is also common to find that the result y for the quick method is only proportional to x, i.e. $y = \beta x$, where β is a constant. The chemist is again fortunate if such a relationship holds exactly. It is likely that his observations will also exhibit a systematic departure α, the same for all x, and a random error with zero mean. Denoting the latter by ε we may write

$$y = \alpha + \beta x + \varepsilon.$$

In general he will not know α, β in advance but will have to estimate them from a series of paired observations of x and y. This is the problem considered in section 2.4, but it is not a solution to the chemist's ultimate problem. He now wishes to *predict* future values of a (mathematical) variable x having observed values of a (random) variate Y. He may be prepared, if σ^2 is small to say, after having obtained estimates $\hat{\alpha}$, $\hat{\beta}$ of α, β, that for his purposes

$$y = \hat{\alpha} + \hat{\beta} x$$

is an exact relationship, whence

$$x = (y - \hat{\alpha})/\hat{\beta}.$$

If ε cannot be ignored he can only say that for a given value of x, the mean value of Y is *estimated* by

$$E(Y) = \hat{\alpha} + \hat{\beta} x.$$

When making use of the quick method if he observes a y value and wants to *estimate* the corresponding true x, it is neither a logically nor a statistically trivial problem to do so, and we postpone a detailed discussion to section 7.2.

17

In the above example several assumptions were made. In particular

(i) observations of x were assumed to be error free;

(ii) it was assumed that, apart from random fluctuations with zero means (the ε), y was a linear function of x;

(iii) the random fluctuations were assumed to have the same normal distribution, irrespective of the value of x and to be independent from observation to observation.

In practical situations it is probably the exception rather than the rule for (i) to (iii) to hold exactly. It is seldom that a relationship between observations is exactly linear, apart from purely random fluctuations; the relationship may be slightly parabolic, for example, tending perhaps to give higher y values for large x than those forecast by the best fitting linear relationship. This type of behaviour may occur, for instance, in a problem like our chemical calibration example if high concentrations of the material raise the temperature in the apparatus and this affects the functioning of the recording instruments in a systematic way that is proportional to x^2, say. In accepting a linear relationship when something like this is happening we are essentially saying that such curvilinear effects are small compared to random fluctuations which are often attributable to unknown or uncontrollable defects in the experimental layout. Small changes in voltage in the mains electricity supply to a circuit, uneven distribution of the constituents in a suspension, temperature variations in the laboratory, frictional effects in the recording apparatus, etc., are examples of such defects.

There are many situations where assumption (iii) will not hold; the method may, for instance, tend to give bigger random errors for large x, or if the instrument gives an unduly high reading for one observation it may give an unduly low reading for the next, i.e. errors may be correlated between successive observations. Some of these complications are dealt with in chapter 4.

The situation where true values x are not obtainable but only values $x' = x + \delta$, where δ is a random variate representing error, is dealt with in chapter 3.

In the chemical calibration example we would, as already stated, ideally like ε to be identically zero so that the (random) variate Y could be replaced by a variable y identically equal to $\alpha + \beta x$. The fact that we have to introduce a random variate to allow for error implies that these errors are superimposed on what Ehrenberg (1968) has termed a *law-like relationship*. His terminology is useful, although his general approach to the problem is very different from ours. Questions about the nature of departures from underlying relationships are of particular relevance to problems of functional relationship discussed in chapter 3.

2.6. Some practical aspects of least squares regression

In this section we discuss some miscellaneous practical aspects of fitting regression lines to observational data when least squares is appropriate. We assume x is a variable measured without error. Therefore we do not speak of the variance of x as we would if it were a variate; we shall however refer to the sum of squares of x and so on, applying this term in an analogous way to that used for observed (random) variates.

Many writers call x the *independent* variable and y the dependent variate in line with the mathematical concept of independent and dependent variables. In bivariate regression, where X, like Y, is a variate, it is still sometimes called an *independent* variate. An essential feature of any non-trivial bivariate regression is that X, Y shall *not* be independent in the statistical sense, so the terminology is confusing. The confusion becomes even more marked in multiple regression. We adopt here one of several alternative terminologies that have been suggested and continue to refer to Y as the dependent variate, but call X a regressor variate and x a regressor variable, or more simply a *regressor* if we do not wish to distinguish between a variate and a variable. Note that in bivariate regression, if we regress X on Y, the roles and names of the variates are interchanged, Y becoming the regressor. This terminology is quite widely used (see, e.g., Cox (1968)).

The various uses of regression are discussed in detail in chapter 7; there are clearly circumstances when regression lines provide a useful summary of data that may sometimes be used in further studies, e.g. regression lines corresponding to different treatments may be tested for parallelism or identity.

Prediction is perhaps one of the commonest uses of a regression equation; the basic problem here is, given a sample of values (x_i, y_i), the regression of Y on x being linear, what can be *predicted* about values of Y corresponding to a further observed x value? This problem involves specific assumptions about the distribution of Y and it is discussed again at various points in this book.

In this section we confine ourselves to one regression line, leaving to chapter 7 questions such as tests for parallelism. We consider now such questions as the properties of estimators $\hat{\alpha}$, $\hat{\beta}$ and the consistency of data with a linear regression hypothesis. We quote below some relevant results; the derivations are given in most standard texts.

If $\hat{\alpha}$, $\hat{\beta}$ are given by (2.5) and (2.6), then the estimated variance of $\hat{\beta}$ is given by

$$\mathrm{Var}(\hat{\beta}) = \hat{\sigma}^2/S_{xx} \tag{2.10}$$

where $\hat{\sigma}^2$ is given by (2.8) and

$$S_{xx} = \sum x_i^2 - (\sum x_i)^2/n,$$

is the sum of squares for x.

If we denote by $\hat{y}_k$ the estimated mean of Y when $x = x_k$, i.e.

$$\hat{y}_k = \hat{\alpha} + \hat{\beta} x_k$$

then

$$\mathrm{Var}(\hat{y}_k) = \hat{\sigma}^2 \left\{ \frac{1}{n} + \frac{(x_k - \bar{x})^2}{S_{xx}} \right\} \tag{2.11}$$

where $\bar{x}$ denotes the mean of the n observed values of x. Note that x_k need not equal one of the observed x values used in computing $\hat{\alpha}$, $\hat{\beta}$.

If $x_k = 0$, (2.11) gives, since $\hat{y}_k$ then equals $\hat{\alpha}$,

$$\mathrm{Var}(\hat{\alpha}) = \hat{\sigma}^2(1/n + \bar{x}^2/S_{xx}). \tag{2.12}$$

Note that $\hat{\alpha}$, $\hat{\beta}$ are correlated and

$$\mathrm{Cov}(\hat{\alpha}, \hat{\beta}) = -\hat{\sigma}^2 \bar{x}/S_{xx}. \tag{2.13}$$

From (2.5) it follows that the least squares regression equation may be written

$$y = \bar{y} + \hat{\beta}(x - \bar{x}) \tag{2.14}$$

where $\bar{y}$ and $\hat{\beta}$ are independent. Using this fact the result (2.11) follows immediately from (2.14) and (2.10).

If the regression equation based on n pairs of observations is used to *predict* the value y_{n+1} corresponding to a further observed x value, x_{n+1}, some care is needed in determining the variance of the predicted value $\hat{y}_{n+1}$. The true value of y_{n+1} will be

$$y_{n+1} = \alpha + \beta x_{n+1} + \varepsilon_{n+1}. \tag{2.15}$$

In predicting y_{n+1} we estimate α, β by $\hat{\alpha}$, $\hat{\beta}$, and an unbiased point estimator of y_{n+1} is $\hat{\alpha} + \hat{\beta}x_{n+1}$ since this has mean value $\alpha + \beta x_{n+1}$ and $E(\varepsilon_{n+1}) = 0$. The variance of $\hat{\alpha} + \hat{\beta}x_{n+1}$ is given by (2.11) with $k = n+1$. Since $\text{Var}(\varepsilon_{n+1}) = \sigma^2$, the variance of a predicted $\hat{y}_{n+1}$ corresponding to a further observed x_{n+1} is estimated by

$$\text{Var}(\hat{y}_{n+1}) = \hat{\sigma}^2\{(n+1)/n + (x_{n+1}-\bar{x})^2/S_{xx}\}. \tag{2.16}$$

Note carefully the difference between (2.11), the variance of an estimated mean, and (2.16). The point estimates of $\hat{y}_k$ and $\hat{y}_{n+1}$ will be identical if $x_k = x_{n+1}$, but they are conceptually different and have different variances.

Some implications of (2.10) to (2.16) are:

(i) Good estimation requires the variances of the estimators to be as small as possible. As $\hat{\sigma}^2$ is an unbiased consistent estimator of σ^2 we may expect it to have a value near σ^2 for not too small values of n. Thus $\text{Var}(\hat{\beta})$ may be reduced by increasing S_{xx}. This can be achieved by making $(x_i - \bar{x})$ numerically large for all i. If there are an even number of observations and for practical reasons these cannot exceed K in modulus, then half the observations should be taken at $x = K$ and half at $x = -K$ to maximize S_{xx}. This would be an optimum procedure if we were sure the regression were truly linear and the errors homoscedastic. In practice we usually want some check upon these assumptions, so we spread our observations over the range of x values of interest. Sometimes the x values are chosen for experimental convenience, or by necessity, rather than on statistical grounds.

(ii) The variances given by (2.11) and (2.16) are clearly a minimum for given n, S_{xx} and $\hat{\sigma}^2$ when $\hat{x}_k$ or $\hat{x}_{n+1} = \bar{x}$ and they increase by an amount proportional to the square of the departures of

$\hat{x}_k$, $\hat{x}_{n+1}$ from $\bar{x}$. This is not unreasonable as it is clear from (2.14) that the estimated line is constrained to pass through the sample means and that departures of $\hat{\beta}$ from β have no effect on estimates of y when $x = \bar{x}$, but that such departures will have an ever increasing effect as x moves away from $\bar{x}$. In general (2.11) and (2.16) will be reduced by increasing n or S_{xx}. Confidence limits for estimates can be obtained by multiplying the appropriate estimated

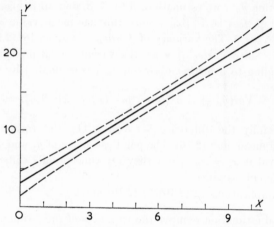

FIG. 2.3. Confidence limits for $E(Y)$ for various X

standard errors (i.e. the square roots of the variances) by suitable t value with $n-2$ degrees of freedom (the degrees of freedom associated with $\hat{\sigma}^2$). For some numerical data given in table 2.1, 95% confidence limits for the estimated mean value of Y corresponding to given x are shown by dotted lines in figure 2.3. The solid line represents the fitted regression line. Confidence limits for the mean of Y corresponding to a given x are

$$\hat{\alpha} + \hat{\beta}x \pm t_{100-\alpha, n-2}\hat{\sigma}\sqrt{\{1/n + (x - \bar{x})^2/S_{xx}\}} \qquad (2.17)$$

where α is the confidence level and $t_{100-\alpha, n-2}$ is the appropriate tabulated value of $|t|$ with $n-2$ degrees of freedom, i.e. the α percentile of $|t|$.

Example 2.1. The following 'observed' values of x and y, (table 2.1) have been obtained by adding random standard normal deviates (Quenouille, 1959) to values of y corresponding to the given x in accordance with the exact relationship $y = 3 + 2x$.

TABLE 2.1. Observed values of a variable x and a variate Y.

x	0	1	2	3	4	5	6	7	8	9	10
y	4·28	2·87	6·62	8·90	11·12	15·31	15·47	17·42	17·93	21·22	21·67

Applying the appropriate least squares formulae we easily obtain $S_{xx} = 110$, and in an analogous notation $S_{yy} = 437\cdot7164$, and $S_{xy} = 215\cdot67$, whence

$$\hat{\beta} = S_{xy}/S_{xx} = 1\cdot9606,$$

$$\hat{\alpha} = 3\cdot0024, \quad \hat{\sigma}^2 = 1\cdot6518.$$

The 95% confidence limits for β are $1\cdot9606 \pm 0\cdot277$, and for the mean of Y corresponding to $x = x_k$ they are given by (2.17) as

$$3\cdot0024 + 1\cdot9606x_k \pm 0\cdot8759\sqrt{\{1 + (x_k - 5)^2/10\}} \qquad (2.18)$$

The limits in table 2.2 have been evaluated from (2.18). These limits and the above regression computations provide the basis of figure 2.3.

TABLE 2.2. The 95% confidence limits for mean Y estimated from a regression equation.

x	Estimated $E(Y)$ and limits.
0	$3\cdot00 \pm 1\cdot64$
1	$4\cdot96 \pm 1\cdot41$
2	$6\cdot92 \pm 1\cdot21$
3	$8\cdot88 \pm 1\cdot04$
4	$10\cdot84 \pm 0\cdot92$
5	$12\cdot81 \pm 0\cdot88$
6	$14\cdot77 \pm 0\cdot92$
7	$16\cdot73 \pm 1\cdot04$
8	$18\cdot09 \pm 1\cdot21$
9	$20\cdot65 \pm 1\cdot41$
10	$22\cdot61 \pm 1\cdot64$

B

From tables 2.1 and 2.2 we note that two of the 'observed' y out of 11 fall outside these limits. This emphasizes the fact that the limits are for means. Confidence limits for individual observations are wider and are based upon the variance (2.16).

Hypothesis tests about β can easily be made using the t-distribution. To test whether $\hat{\beta}$ is consistent with a theoretical value β_0, say, the test statistic is

$$|t| = \frac{|\hat{\beta} - \beta_0|}{\sqrt{\{\text{Var}(\hat{\beta})\}}}$$

where $\text{Var}(\hat{\beta})$ is given by (2.10). Under the hypothesis $\beta = \beta_0$, t has a t distribution with $n-2$ degrees of freedom. With a test of size α (not to be confused with the parameter α), the hypothesis will be accepted for any β_0 lying within the $100-\alpha$ % confidence interval for β.

2.7. Adequacy of the simple least squares model

The model considered in sections 2.4, 2.5 and 2.6 will be referred to for simplicity as the *classical* least squares regression model, or more simply as the classical regression model, to distinguish it from other least squares models to be introduced in later chapters. In section 2.5 we mentioned that breakdown of the model may take the form of (i) systematic trends away from linearity or (ii) failure of the homoscedasity, independence and normality assumptions for the random departures or errors.

Departures from linearity may sometimes be detected by straight-forward plotting of points, when any marked curvature will be detected. Less marked curvature may be detected by plotting of residuals after fitting a straight line by least squares, an approach that is further discussed in section 7.7.

More formal tests for homogeneity of error and departures from linearity are available if, for each x value, there is more than one observed y value. The test for homogeneity of error variance is then carried out by estimating σ^2 separately for each x. Suppose that corresponding to $x = x_i$ the observed values of y are $y_{i_1}, y_{i_2} \ldots, y_{in_i}$, for $i = 1, 2, \ldots, n$. Then, under the hypothesis of homogeneity of random departures or errors and normality

$$s_i^2 = \sum_j \{y_{ij}^2 - (\sum_j y_{ij})^2/n_i\}/(n_i-1)$$

are, for all i, independent unbiased estimators of σ^2, each with $n_i - 1$ degrees of freedom. The hypothesis of homogeneity of errors may then be investigated using a well-known test due to Bartlett (1937). Writing $f_i = n_i - 1$ this test requires the computation of the statistic

$$T = [(\textstyle\sum f_i)\ \ln s^2 - \sum(f_i\ \ln s_i^2)]/C, \quad \text{where}\quad s^2 = (\textstyle\sum f_i s_i^2)/\sum f_i$$

and

$$C = 1 + \frac{1}{3(n-1)}\ [\textstyle\sum(1/f_i) - 1/(\sum f_i)].$$

Under the hypothesis of homogeneity of variance T has a χ^2 distribution with $n-1$ degrees of freedom.

If Bartlett's test provides no reasonable grounds for rejecting the hypothesis of homoscedasity, a test for departures from linearity may be made by partitioning the *between groups* sum of squares for y in a one-way classification analysis of variance. Here each x value defines one group, and the usual analysis of variance partitions the total sum of squares into a between groups sum of squares with $n-1$ degrees of freedom and a within groups sum of squares with $N-n$ degrees of freedom, where $N = \sum n_i$ is the total number of observed y values. Note that if the individual s_i^2 have been computed for Bartlett's test, then the sum of squares within groups is $\sum(f_i s_i^2)$. It is not difficult to show algebraically that the regression sum of squares comprising the last term of (2.7) is a component of the between groups sum of squares with 1 degree of freedom. The remaining sum of squares with $n-2$ degrees of freedom represents deviations from regression. If the *only* source of deviations of the group means from the regression line is that associated with the ε_{ij} the corresponding mean square will have the same expectation σ^2 as the within groups mean square. The usual variance ratio tests may be performed using the analysis of variance indicated in table 2.3.

The ratios $V_1 = M_R/M_W$ and $V_2 = M_D/M_W$ are computed and the F tables entered with 1, $N-n$ and $n-2$, $N-n$ degrees of freedom respectively. A significant value of V_1 indicates that there is a linear component of trend; if V_2 is not significant it may be taken that there is no firm evidence of any other trend; if V_2 is significant there is evidence of a departure from linearity; this departure may be in the form of curvature, or some other non-random departure from linearity such as periodicity.

Failure to detect a significant departure from linearity does not mean the model has been *proved* correct. It may be possible to set up alternative models for which there is also no significant evidence of departure.

TABLE 2.3. Significance of regression and deviations from regression.

Analysis of variance			
Source	Degrees of freedom	Sums of Squares	Mean Squares
Regression	1	$\hat{\beta}S_{xy}$	$M_R = \hat{\beta}S_{xy}$
Deviations from regression	$n-2$	$S_D = S_B - \hat{\beta}S_{xy}$	$M_D = S_D/(n-2)$
Between groups	$n-1$	S_B	
Within groups	$N-n$	$S_W = S_T - S_B$	$M_W = S_W/(N-n)$
Total	$N-1$	S_T	

Cox (1968) gives an example where he suggests that over a limited range of values there may be little evidence of systematic departures from linearity when we consider the relation between x and y, or the relation between $\log x$ and y or between $\log x$ and $\log y$ for the same data. It is possible to compare the goodness of fit of a straight line for the first two, but the third cannot be compared directly with the other two.

There may be sound theoretical grounds for preferring one relationship to another. For example, if we measure the diameter and weight of certain spheres, and believe them all to have more or less the same density, then on theoretical grounds we would expect to get a better linear relationship between weight and the cube of diameter than between weight and diameter.

The relationship between weight and logarithm of diameter would also be expected to be approximately linear with β taking a value near three, and the value of α depending upon the density. In this example we expect linearity on physical rather than statistical grounds.

Finally in this section we note that if the origin is transferred to the mean of the observations, $(\bar{x}, \bar{y})$, then writing

$$x' = x - \bar{x} \text{ and } y' = y - \bar{y},$$

(2.14) becomes $y' = \beta x'$. Thus, as far as estimation of β is concerned, there is no loss of generality if the origin is taken at the mean of the observations. This leads to simplification of some commonly occurring expressions, e.g. $S_{xy} = S_{x'y'} = \sum x_i' y_i'$, etc.

Exercises

2.1 Show that for a bivariate normal distribution

$$\text{Var}(Y | X = x) = \sigma_y^2 (1 - \rho^2) = \sigma^2, \text{ say,}$$

and that given n pairs of observations (x_i, y_i) the maximum likelihood estimator of σ^2 has the same form as (2.7). (Note the implication that σ^2 is independent of x.)

2.2. Obtain an expression for confidence limits for β when an estimate is based on n pairs of observations for the bivariate normal regression model.

2.3. Verify the statement made in section 2.4 that $\hat{\sigma}^2$ given by (2.7) is a biased estimator of σ^2.

2.4. What modifications are needed to the arguments in section 2.4 if a straight line is fitted to pass through the origin (i.e. the line $y = \beta x$ is fitted by least squares)? How are the formulae for variances of the estimators given in section 2.6 modified in this case? How would the analysis of variance in table 2.3 and related tests be modified in this case?

2.5. A classical least squares regression line is fitted to some data. How would you test whether the hypothesis that the true regression line passed through the origin was reasonable?

2.6. How would you test whether a classical least squares regression line was consistent with α having a value α_0?

2.7. How would you test whether the unbiased estimator $\hat{\sigma}^2$ in (2.8) was consistent with a value σ_0^2 for σ^2? (Hint: consider the distribution of $\hat{\sigma}^2/\sigma_0^2$ under the hypothesis.)

2.8. If X, Y are uniformly distributed over a parallelogram with vertices at $(0, a)$, (c, d), $(0, -a)$, $(-c, -d)$ where a, c, d are positive and unequal, obtain the regressions of Y on X and X on Y, establishing that one of these is a straight line, and the other consists of three segments of straight lines joined together.

CHAPTER 3

Law-like Relationships in the Presence of Random Variation

3.1. Functional and structural relationships

The type of relationships considered in detail in the previous chapter involved expressions for the mean of the dependent variate as a linear function of the regressor. It was true that in certain practical situations our hope would be that for a given x, $E(Y)$ would have small variance. For example, in the chemical calibration problem introduced in section 2.5 one would hope that ε was always small.

In practice engineers, physicists, chemists, biologists, industrialists, psychologists, economists and others often face a situation in which they believe, or hope, there is a basic mathematical relationship between variables with which their data would accord were it not for the fact that this relationship is obscured to some extent by 'random' fluctuations, perhaps associated with both variables.

As we pointed out in section 2.5, these fluctuations, which in general we shall call departures, very often represent genuine variability in the experimental material rather than errors of measurement. This is particularly true in most biological or economic studies. In this case the underlying relationship may be looked upon as representing an average relationship for the population under study. Departures from the relationship represent what may be termed individuality. The average relationship is a useful concept in comparing populations. For two or more populations we may well ask whether the average relationships differ.

For example, suppose arm length and leg length are measured on a random sample of adult British males and the paired observations are plotted. These will be found to lie on a smooth curve not very different from a straight line, and the approximation to a straight line will be even better if we plot logarithms of the observations. A biologist may find he gets very similar lines for similar data for samples of Frenchmen, Americans, Indians and Nigerians, but that for a sample of Australian aborigines his fitted line appears to have a different slope. He will want to know whether this is a real difference, or whether it can be attributed to sampling variation.

To take another example, linear relationships have been found to hold approximately between measurements of the amount of fruit (reproductive growth) and vegetative growth of plants. The precise relationship varies from species to species and this is of interest to plant physiologists as many of their investigations are directed towards upsetting this balance so that a plant will devote a bigger proportion of its energies to producing edible crop.

Nature herself has altered relationships between sizes of parts of organisms in the evolutionary process, and these changes can be studied. We might find a linear relationship, perhaps after taking logarithms, to hold fairly well between measurements of height above the ground of the top of dogs' heads and their body length, nose to tail, more or less irrespective of their breed or lack of breed; but there would be notable exceptions such as dachshunds. Suppose, however, that we confined our attention entirely to dachshunds; we would probably find that *within* that breed a linear relationship held between the two measurements, apart of course from superimposed random variability representing individuality. This relationship would be different from that holding for more conventional breeds; it is also possible that we may find *within* other breeds linear relationships, sometimes similar to that holding in general *between* breeds (other than exceptional ones) but sometimes rather different. This example indicates an important point of principle – namely that some of the departure variability associated with a relationship may be due to differences in the relationship as we move from group to group (groups corresponding to distinct breeds in dogs). This principle features prominently in the published discussion on the paper by Ehrenberg (1968). Differences in the relationships within groups may only be clearly observable when there is extreme departure from the between groups pattern, but the concept of two types of relationship – those holding within and those holding between groups – is important. Between and within groups variation is an essential concept of various aspects of canonical analysis which is widely used in problems of discrimination, where the object is to allocate individuals to groups on the basis of sets of measurements made upon them. Canonical analysis is discussed again in section 6.5.

Law-like relationships have a predictive role in calibration experiments in situations where departures from them are negligible, but they are of greater importance for the insight they may give into the way a process is operating; they can often be looked upon as telling us how x and y are related on average, or under idealized circumstances.

A model with an intuitive appeal to biologists and economists is one that supposes there is an underlying relationship between mathematical variables ξ, η of the form

$$\eta = \alpha + \beta\xi. \tag{3.1}$$

When there are no departures any such line is uniquely determined if two points (ξ_1, η_1) and (ξ_2, η_2) are known to lie on it. When, however, observations are subject to additive random variation, the observed values (x_i, y_i), $i = 1, 2, \ldots, n$ may be written

$$x_i = \xi_i + \delta_i, \tag{3.2}$$

$$y_i = \eta_i + \varepsilon_i = \alpha + \beta\xi_i + \varepsilon_i. \tag{3.3}$$

The unknowns are α, β the ξ_i and δ_i and ε_i. Note that if we could observe ξ_i exactly, which would imply $\delta_i \equiv 0$, and $x_i = \xi_i$, then (3.3) would take the form

$$y_i = \alpha + \beta x_i + \varepsilon_i \tag{3.4}$$

and with the further assumption that the ε_i are independent of one another and $N(0, \sigma^2)$ for all i the model is formally the same as that for classical least squares regression. The model in (3.2) and (3.3) has sometimes been called the model for *regression with both variables subject to error*. This is extending Galton's concept of regression even further than was done in chapters 1 and 2, and it is of course sometimes misleading to refer to random variation as error – unless one implies by this some sort of error in the model. We prefer to call the underlying law-like relationship between mathematical variables a functional relationship. Given observations

31

that are not true variable values, but random variates decomposing after the manner of (3.2) and (3.3), the first problem is to estimate α, β in the functional relationship (3.1).

As well as the type of relationship just described it is possible to have relationships between random variates that are themselves obscured by further additive random variability. The underlying relationship may be of the bivariate regression type, or there may be an exact relationship between the random variates; in this latter case it is convenient to speak of the underlying relationship as a *structural* relationship. The distinction we have made between a functional and a structural relationship is not universal in the literature.

Many users of classical regression techniques have applied them when there is an additive random component δ in the regressor. To study the effect of ignoring such a component let us consider the relationship between child height and mid-parent height discussed in section 2.3, but suppose now that instead of the true heights (x_i, y_i) being recorded, random errors are made in measuring these so that what is recorded is a pair of values (x_i', y_i') where

$$x_i' = x_i + \delta_i, \tag{3.5}$$

$$y_i' = y_i + \varepsilon_i. \tag{3.6}$$

We suppose δ_i, ε_i to be independent of one another and for different values of i; we assume that for all i they have the same distributions with zero means and variances $\sigma_\delta{}^2$, $\sigma_\varepsilon{}^2$ respectively. A normality assumption is not required for δ_i, ε_i at this stage, but they must be uncorrelated with x_i, y_i. If we assume x_i, y_i to have a joint normal distribution and proceed according to the methods of section 2.2 to estimate the regression of Y on X ignoring δ_i, ε_i (i.e. treating x_i', y_i' as though they were x_i, y_i) we estimate β as follows, where for simplicity we assume the origin is taken at the mean of the observations; we have pointed out in section 2.7 that this results in no loss of generality. Writing b_1 for this estimator of β we have

$$b_1 = \frac{\sum x_i' y_i'}{\sum x_i'^2} = \frac{\sum (x_i + \delta_i)(y_i + \varepsilon_i)}{\sum (x_i + \delta_i)^2}$$

and since we have assumed $\text{Cov}(\delta_i, \varepsilon_i) = 0$ and x_i, y_i are independent of δ_i and ε_i it follows that

$$E(b_1) = \frac{\text{Cov}(X, Y)}{\sigma_X^2 + \sigma_\delta^2} \,.$$

The true regression coefficient of Y on X is

$$\beta = \frac{\text{Cov}(X, Y)}{\sigma_X^2}$$

whence it follows that b_1 is a biased estimator of β. Its mean value is always nearer to zero than that of β. This effect is sometimes called the *attenuation* effect of *errors* in X. Note that $E(b_1)$ does not involve σ_ε^2 and errors in the observed values of Y do not introduce bias into the usual regression estimator. The estimated regression of X on Y will be affected by errors in Y.

Lindley (1947) has shown that this is not the whole story, for even if the true regression of Y on X is linear (as it is for the bivariate normal distribution) it does not follow that the regression of Y' on X' is linear. The condition under which it will be linear if that of Y on X is linear is that the cumulant generating function of X is a multiple of the cumulant generating function of δ. A proof was first given by Lindley (1947), and one is also given by Kendall and Stuart (1967, chapter 29). Linearity of regression is thus maintained in the example we have been considering if we add a further assumption of normality for the δ_i, since the cumulant generating function of the δ_i is then $\sigma_\delta^2 t^2/2$ and that of X is $\sigma_X^2 t^2/2$, whence the requirement of the theorem is met.

We shall discuss the estimation of the parameters in a structural relationship more fully in section 3.5.

3.2. Comments on functional relationships

We now consider situations where we postulate an underlying linear relationship between a pair of mathematical variables. The parameters are to be estimated from observations containing an additive stochastic or random element. There is an extensive literature on this topic including Lindley (1947), Tukey (1951), Madansky (1959), Williams (1959, chapter 11), Sprent (1966),

Carlson, Sobel and Watson (1966), Kendall and Stuart (1967, chapter 29) and Lindley and El-Sayyad (1968).

In this chapter we only consider cases where δ_i, ε_i in (3.2) and (3.3) have very simple properties. We shall assume they have the same normal distribution for all i, are independent of ξ_i, η_i, that they are independent from observation to observation and have zero means. Lindley (1947) showed that in these circumstances there is no loss of generality if we put $\alpha = 0$, and that this can be achieved by taking our origin at the mean of the observations. We shall assume this is the case. In much of the literature a further assumption is made that δ_i, ε_i are not correlated with each other. Although we shall make this assumption for the present, it is often unrealistic in practice and dropping it results only in minor algebraic complications. We consider this matter further in sections 3.3. and 3.4. We write $\sigma_\delta{}^2$, $\sigma_\varepsilon{}^2$ for the variances of δ_i, ε_i.

A notable feature of the regression situations in chapter 2 was that we could proceed to a point estimate of β without any knowledge of σ^2, although an estimate, at least, of σ^2 was required for interval estimation. If we write down the likelihood function in the present case we find that $L^* = \ln L$ is equal to

$$-\frac{n}{2} \ln \sigma_\delta{}^2\sigma_\varepsilon{}^2 - \frac{1}{2} \frac{\sum(x_i - \xi_i)^2}{\sigma_\delta{}^2} - \frac{1}{2} \frac{\sum(y_i - \beta\xi_i)^2}{\sigma_\varepsilon{}^2} + k \qquad (3.7)$$

where k is a constant. Although estimation of β is usually of prime interest, without further assumptions $\sigma_\delta{}^2$, $\sigma_\varepsilon{}^2$ and the n values of the ξ_i are all unknown. The $n+3$ normal equations are obtained in the usual way by differentiating (3.7) and setting the derivatives to zero. They are

$$\sum(y_i - \beta\xi_i)\xi_i = 0, \qquad (3.8)$$

$$\sum(x_i - \xi_i)^2 = n\sigma_\delta{}^2, \qquad (3.9)$$

$$\sum(y_i - \beta\xi_i)^2 = n\sigma_\varepsilon{}^2, \qquad (3.10)$$

$$\sigma_\varepsilon{}^2(x_i - \xi_i) + \beta\sigma_\delta{}^2(y_i - \beta\xi_i) = 0, \, i = 1, 2, \ldots, n. \qquad (3.11)$$

From (3.11) we get

$$\xi_i = \frac{x_i\hat{\sigma}_\varepsilon{}^2 + y_i\beta\hat{\sigma}_\delta{}^2}{\hat{\sigma}_\varepsilon{}^2 + \hat{\beta}^2\hat{\sigma}_\delta{}^2} \qquad (3.12)$$

where the circumflex denotes a maximum likelihood estimator. Equation (3.12) implies

$$x_i - \hat{\xi}_i = \frac{(\hat{\beta}x_i - y_i)\hat{\beta}\hat{\sigma}_\delta{}^2}{\hat{\sigma}_\varepsilon{}^2 + \hat{\beta}^2\hat{\sigma}_\delta{}^2} \tag{3.13}$$

and

$$y_i - \hat{\beta}\hat{\xi}_i = \frac{(y_i - \hat{\beta}x_i)\hat{\sigma}_\varepsilon{}^2}{\hat{\sigma}_\varepsilon{}^2 + \hat{\beta}^2\hat{\sigma}_\delta{}^2} \tag{3.14}$$

and substitution of (3.13) and (3.14) in (3.9) and (3.10) gives

$$\hat{\beta}^2 = \hat{\sigma}_\varepsilon{}^2 / \hat{\sigma}_\delta{}^2, \tag{3.15}$$

a result that implies that the maximum likelihood estimator of the slope is either plus or minus the square root of the ratio of the estimators of the variances of ε, δ.

The result (3.15), first obtained by Lindley (1947), is generally taken to mean that maximum likelihood estimation breaks down in this case. The difficulty however is inherent in the problem as stated, which has no satisfactory solution.

Kendall and Stuart (1967) show by a geometric heuristic argument where the difficulty lies. Their argument follows these lines. Each observation (x_i, y_i) emanates from an *unknown* true point (ξ_i, η_i). *If* we knew $\sigma_\delta{}^2$ and $\sigma_\varepsilon{}^2$ we could draw elliptic confidence regions for ξ_i, η_i appropriate to any desired probability level and centred at (x_i, y_i). Heuristically, the problem of estimating β is to find a line that cuts as many of these ellipses as possible. Indeed, if we constructed, say, 95% confidence ellipses, we would certainly not be happy about accepting a hypothesis about a linear relationship unless a line could be found to cut a very high proportion of such ellipses and passed fairly close to the remainder. If we have no information about $\sigma_\delta{}^2$ and $\sigma_\varepsilon{}^2$ not only are we without knowledge of the geometric area of the confidence regions, but we do not even know the eccentricity of the ellipses. A knowledge of the ratio of these variances will determine the eccentricity of the ellipses, but this in itself will not tell us whether a straight line fits the data as it will not tell us the area of the ellipses. This is clear from figure 3.1, which shows three groupings of identical points. Figure 3.1(*a*) shows the points without any information about confidence regions. Ellipses of different eccentricity are shown in figures 3.1(*b*) and 3.1(*c*), and in each case we suppose these ellipses to represent 95%

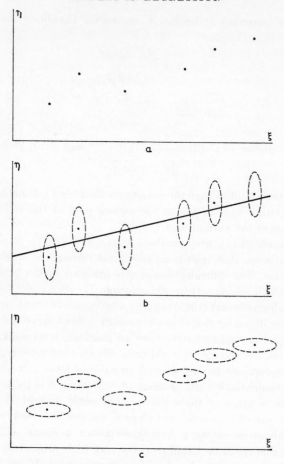

FIG. 3.1 (a–c). Heuristic argument for the need for
information on departure variances when fitting a
straight line

confidence regions. Although in both cases we know the ratio of
the variances, $\lambda = \sigma_\varepsilon^2/\sigma_\delta^2$ as we know the eccentricity, it is clear
that for the situation in figure 3.1(b) the hypothesis of a straight line
would be acceptable, and equally clear that it would not be acceptable for the situation depicted in figure 3.1(c). This diagram shows
questions of existence and slope of any line to be intimately bound
up with a knowledge of σ_δ^2 and σ_ε^2. In section 3.3 we shall see that

36

maximum likelihood behaves in a way that is not so unreasonable as (3.15) suggests at first sight, nevertheless it does not provide a satisfactory method of fitting.

A certain amount of cynical comment has been made from time to time about the ability of the practical man, given a pencil and ruler, to fit a straight line to data of the type in figure 3.1(a) by eye when the statistician says it cannot be done. The heuristic argument given above makes it clear that in fitting such a line the practical man is assuming something about departures – essentially that the points could have arisen from a linear relationship with superimposed additive random departures.

Although maximum likelihood breaks down with no knowledge about σ_δ^2, σ_ε^2 it is often possible to do something statistically, using, for example, the method of Wald (1940) or the related method given by Bartlett (1949).

Lindley and El-Sayyad (1968) have discussed this problem from a Bayesian viewpoint, and they found that with only vague prior distributions for the variances the solution ran into difficulties of another kind. We shall see in section 3.3 that maximum likelihood estimators are inconsistent without knowledge of departure vari-ances, i.e. they converge to incorrect values. The Bayesian solution, on the other hand, does not converge, i.e. it gives an estimator with finite non-zero dispersion however large the sample may be. This would seem to reflect the same fundamental difficulty of lack of identifiability of the parameters without further knowledge. Some may prefer an estimator that gives only a range of possible values for the parameter, to an estimator known to converge to a value which, although it may not be very different from the true value, will not in general be equal to it. We discuss the more satisfactory situation where we have some knowledge about σ_δ^2 and σ_ε^2 in section 3.4. In section 3.3 we explore more fully the behaviour of maximum likelihood estimators without such information.

3.3. Behaviour of the maximum likelihood estimators with unknown variances

Equation (3.15) reflects the fact that we cannot disentangle the estimators $\hat{\beta}$, $\hat{\sigma}_\delta^2$ and $\hat{\sigma}_\varepsilon^2$. Substitution from (3.12) and (3.14) in (3.8) using (3.15) gives

$$\hat{\beta}^2 = \left(\sum y_i^2\right)/\left(\sum x_i^2\right). \tag{3.16}$$

There is a choice of sign for $\hat{\beta}$, and the obvious choice is to take the sign the same as that of $\sum x_i y_i$. Now let us suppose that $\sum \xi_i^2/n$ converges to V, say. This may prove a quite reasonable assumption in practice, and it implies that, with the independence assumptions we have made, $\sum y_i^2/n$ converges to $\beta^2 V + \sigma_\varepsilon^2$ and $\sum x_i^2/n$ converges to $V + \sigma_\delta^2$, and thus

$$\hat{\beta}^2 \to \frac{\beta^2 V + \sigma_\varepsilon^2}{V + \sigma_\delta^2},$$

a result given by Lindley and El-Sayyad (1968). Thus $\hat{\beta}^2$ converges to β^2 only if $\beta^2 = \sigma_\varepsilon^2/\sigma_\delta^2$ or $\sigma_\varepsilon^2 = \sigma_\delta^2 = 0$, and we may exclude the latter case as there is then no statistical problem, and in fact the likelihood function could no longer be written down in the form (3.7). If there is a wide spread of observed points and departures are not too large, V will be large compared to σ_δ^2, and if, in addition, β^2 is not too small, $\beta^2 V$ will be large compared to σ_ε^2 and the inconsistency of $\hat{\beta}$ as an estimator of β will not be large. Some may feel that this is a trivial result in the sense that it tells us something about the position when errors are small, but gives no guarantee that the errors are in fact small in any particular case; there are also strong reasons for objecting to an estimator that converges to a value that we know not to be the true value with probability one; but the above argument at least shows the estimator to be more sensible than a cursory glance at (3.15) might suggest.

When $\hat{\beta}$ is small or the departures large, $\hat{\beta}$ may be markedly inconsistent. This is in accordance with reality, as it is in these circumstances that our practical man will have difficulty in obtaining a good fit by eye, and indeed he should not be attempting to do so. Equation (3.15) implies that $\hat{\sigma}_\delta^2$, $\hat{\sigma}_\varepsilon^2$ are unlikely to be consistent estimators. Using (3.8) to (3.11) it is possible to show that

$$\hat{\sigma}_\delta^2 = \frac{1}{2n}\left\{\sum x_i^2 - (\sum x_i^2/\sum y_i^2)^{1/2}\left|\sum x_i y_i\right|\right\} \tag{3.17}$$

$$\hat{\sigma}_\varepsilon^2 = \frac{1}{2n}\left\{\sum y_i^2 - (\sum y_i^2/\sum x_i^2)^{1/2}\left|\sum x_i y_i\right|\right\}$$

$$= \frac{1}{2n}\left\{\sum y_i^2 - \hat{\beta}\sum x_i y_i\right\} \tag{3.18}$$

and $\qquad \hat{\xi}_i = (x_i + y_i/\hat{\beta})/2. \tag{3.19}$

38

Note the formal resemblance, apart from the factor 1/2, between (3.18) and (2.7), remembering that (3.18) was obtained taking the origin at the mean.

The consistency of these estimators can be explored under similar assumptions to those made in investigating the consistency of $\hat{\beta}$, and $\hat{\sigma}_\delta{}^2$, $\hat{\sigma}_\varepsilon{}^2$ turn out not to be consistent even after adjustment of the factor $2n$.

It is of some interest to note that in the above situation the estimators are unaltered if we replace the assumption that $\text{Cov}(\delta_i, \varepsilon_i) = 0$, by the assumption that $\text{Cov}(\delta_i, \varepsilon_i) = k$, fixed, providing $k^2 \neq \sigma_\delta{}^2\sigma_\varepsilon{}^2$. Although (3.7) is altered and the algebra is tedious there are no complications. When the errors are perfectly correlated the likelihood function becomes infinite. If $\text{Cov}(\delta_i, \varepsilon_i)$ is taken to be unknown, the likelihood approach runs into serious difficulties as the estimators include a condition $\hat{\sigma}_\delta{}^2\hat{\sigma}_\varepsilon{}^2 = \hat{\sigma}_{\delta\varepsilon}$, where $\hat{\sigma}_{\delta\varepsilon}$ is an estimator of $\text{Cov}(\delta_i,\varepsilon_i)$. The author has found no way of resolving resulting difficulties of estimation.

The estimator $\hat{\xi}_i$ given by (3.19) has a plausible form, but is usually of little interest. The ξ_i have been very appropriately termed *incidental parameters* by Neyman and Scott (1951). The spread of ξ_i values is of some importance in determining the accuracy of an estimate of β, but their individual values are seldom of interest.

3.4. Functional relationships when there is information on departure variance

We retain the assumptions made in section 3.2 about the distribution of δ_i and ε_i, including for the moment the assumption that they are uncorrelated with one another. We make, however, the additional assumption that the ratio of their variances, $\lambda = \sigma_\varepsilon{}^2/\sigma_\delta{}^2$, is known. Considering the likelihood function L, and putting $L^* = \ln L$ we get

$$L^* = -\frac{n}{2}\ln \lambda\sigma_\delta{}^4 - \frac{1}{2}\frac{\sum(x_i-\xi_i)^2}{\sigma_\delta{}^2} - \frac{1}{2}\frac{\sum(y_i-\beta\xi_i)^2}{\lambda\sigma_\delta{}^2} + k$$

where $\sigma_\delta{}^2$, β and the ξ_i are to be estimated. The normal equation for $\hat{\beta}$ is the same as (3.8) and that for $\hat{\xi}_i$ has a similar form to (3.11) except that $\sigma_\varepsilon{}^2$ is replaced by $\lambda\sigma_\delta{}^2$, whence, if $\sigma_\delta{}^2 \neq 0$,

$$\hat{\xi}_i = (\lambda x_i + \hat{\beta}y_i)/(\lambda + \hat{\beta}^2). \tag{3.20}$$

Substitution in (3.8) gives a quadratic in $\hat{\beta}$ with roots

$$\hat{\beta} = \frac{\sum y_i^2 - \lambda\sum x_i^2 \pm \sqrt{\{(\sum y_i^2 - \lambda\sum x_i^2)^2 + 4\lambda(\sum x_i y_i)^2\}}}{2\sum x_i y_i} . \quad (3.21)$$

It is not difficult to show that the likelihood is maximized by taking the optional sign in (3.21) as positive, and this ensures that $\hat{\beta}$ always has the same sign as $\sum x_i y_i$. This is a well known result. If we know not only the ratios, but the actual values of σ_δ^2 and σ_ε^2 there is no difficulty but a little care is needed in handling the normal equations (see Barnett (1967)).

In most practical situations we are unlikely to know λ unless we know σ_δ^2 and σ_ε^2. If we have replicated observations it may be possible, as we shall see below, to obtain independent estimates of these. If this cannot be done there may be theoretical grounds for assuming a value of λ. It may be reasonable, for instance, to assume that departures for both variates have roughly the same distribution in which case it would be not unreasonable to take $\lambda = 1$.

We have already pointed out that the assumption $\text{Cov}(\varepsilon_i, \delta_i) = 0$, may not be realistic. We consider a biological situation in which this assumption seems unreasonable.

Suppose x_i, y_i are the lengths of two limbs for an animal i. For a given species these lengths may be approximately linearly related over all animals, apart from random departures attributable to individuality. Suppose that for animal i the departure δ_i of x_i from ξ_i is positive; it is then likely that the departure ε_i of y_i from η_i will also be positive; similarly negative δ_i is likely to be associated with negative ε_i. This correlation between departures reflects the fact that similar genetical and environmental factors will influence the growth of both limbs. On the other hand it is easy to envisage biological systems in which excessive growth in one organ may be compensated for by reduced growth in another, giving rise to negative correlations between δ_i and ε_i.

At this stage we introduce a change in notation that not only gives us a simple notation for $\text{Cov}(\delta_i, \varepsilon_i)$, but will also be more convenient in the multivariate situations considered in later chapters. We replace σ_δ^2, σ_ε^2 by $\sigma_{\delta\delta}$, $\sigma_{\varepsilon\varepsilon}$ and write $\text{Cov}(\delta_i, \varepsilon_i) = \sigma_{\delta\varepsilon} = \sigma_{\varepsilon\delta}$. This notation for variances and covariances is common in multivariate analysis. Care must be taken not to confuse the *variance* $\sigma_{\varepsilon\varepsilon}$ with a *standard deviation* σ_ε (corresponding to the notation σ_ε^2 for variance). Using this notation and permitting correlations between departures, for given $\sigma_{\delta\delta}$, $\sigma_{\varepsilon\varepsilon}$ and $\sigma_{\delta\varepsilon}$, the logarithm of the likelihood, L^* is given by

$$L^* = -\frac{n}{2}\ln\left(\sigma_{\delta\delta}\sigma_{\varepsilon\varepsilon}-\sigma_{\delta\varepsilon}{}^2\right) - \frac{1}{2(\sigma_{\delta\delta}\sigma_{\varepsilon\varepsilon}-\sigma_{\delta\varepsilon}{}^2)}\left[\sigma_{\varepsilon\varepsilon}\sum(x_i-\xi_i)^2 - \right.$$

$$\left. -2\sigma_{\delta\varepsilon}\sum(x_i-\xi_i)(y_i-\beta\xi_i)+\sigma_{\delta\delta}\sum(y_i-\beta\xi_i)^2\right]+k$$

where k is a constant and it is assumed that $\sigma_{\delta\delta}\sigma_{\varepsilon\varepsilon} \neq \sigma_{\delta\varepsilon}{}^2$, i.e. that the errors are not perfectly correlated. The equations for estimation of β and ξ_i are

$$\sum(y_i-\beta\xi_i)\xi_i\sigma_{\delta\delta}-\sum\sigma_{\delta\varepsilon}\xi_i(x_i-\xi_i) = 0, \qquad (3.22)$$

and

$$\sigma_{\varepsilon\varepsilon}(x_i-\xi_i)-\sigma_{\delta\varepsilon}\{\beta(x_i-\xi_i)+y_i-\beta\xi_i\}+\beta\sigma_{\delta\delta}(y_i-\beta\xi_i) = 0,$$
$$i = 1, 2, \ldots, n. \qquad (3.23)$$

It is possible to eliminate ξ_i from these equations and hence to obtain $\hat{\beta}$. The algebra is tedious and a better procedure is as follows. From (3.23) we easily obtain

$$y_i-\beta\xi_i = (y_i-\beta x_i)P/R, \qquad (3.24)$$

and

$$x_i-\xi_i = (y_i-\beta x_i)Q/R, \qquad (3.25)$$

where $P = \sigma_{\varepsilon\varepsilon}-\beta\sigma_{\varepsilon\delta}$, $Q = \sigma_{\varepsilon\delta}-\beta\sigma_{\delta\delta}$, $R = \sigma_{\varepsilon\varepsilon}-2\beta\sigma_{\varepsilon\delta}+\beta^2\sigma_{\delta\delta}$; P, Q R, do not involve ξ_i.

Effectively maximization of the likelihood only requires minimization of the term in square brackets in L^*, i.e. minimization of

$$V = \sigma_{\varepsilon\varepsilon}\sum(x_i-\xi_i)^2-2\sigma_{\varepsilon\delta}\sum(x_i-\xi_i)(y_i-\beta\xi_i)+\sigma_{\delta\delta}\sum(y_i-\beta\xi_i)^2. \quad (3.26)$$

On substituting from (3.24) and (3.25) in (3.26) we find that V is proportional to

$$U = \frac{\sum(y_i-\beta x_i)^2}{R}.$$

Thus determination of $\hat{\beta}$ involves a principle of weighted least squares. U can conveniently be written

$$U = \frac{b_{yy} - 2\beta b_{xy} + \beta^2 b_{xx}}{\sigma_{\varepsilon\varepsilon} - 2\beta\sigma_{\varepsilon\delta} + \beta^2\sigma_{\delta\delta}} \tag{3.27}$$

where $\qquad b_{yy} = \sum y_i{}^2, \ b_{xy} = \sum x_i y_i, \ b_{xx} = \sum x_i{}^2,$

remembering that the origin has been chosen so that $\bar{x} = \bar{y} = 0$.

Equation (3.27) was used by Sprent (1966) as a starting point for considering the estimation of β in this and more complicated situations. It is convenient to write

$$\mathbf{B} = \begin{pmatrix} b_{xx} & b_{xy} \\ b_{xy} & b_{yy} \end{pmatrix}$$

and

$$\mathbf{W} = \begin{pmatrix} \sigma_{\delta\delta} & \sigma_{\delta\varepsilon} \\ \sigma_{\delta\varepsilon} & \sigma_{\varepsilon\varepsilon} \end{pmatrix}$$

In Sprent (1966) generalized least squares was used as a principle in its own right, but the above demonstration of its equivalence to maximum likelihood strengthens the case for using the principle, and provides an even more direct justification than the novel and interesting Bayesian argument proposed by Lindley in the published discussion on that paper.

The minimization of U is most easily performed by noting that both numerator and denominator in (3.27) are quadratic forms, say, Q_1, Q_2 respectively. If λ is chosen so that $Q_1 - \lambda Q_2$ is a perfect square, then

$$Q_1 - \lambda Q_2 = (b_{xx} - \lambda\sigma_{\delta\delta})\left\{ \beta^2 - \frac{2\beta(b_{xy} - \lambda\sigma_{\delta\varepsilon})}{b_{xx} - \lambda\sigma_{\delta\delta}} + \frac{b_{yy} - \lambda\sigma_{\varepsilon\varepsilon}}{b_{xx} - \lambda\sigma_{\delta\delta}} \right\}$$

$$= (b_{xx} - \lambda\sigma_{\delta\delta})(\beta - \hat{\beta})^2$$

where

$$\hat{\beta} = \frac{b_{xy} - \lambda\sigma_{\varepsilon\delta}}{b_{xx} - \lambda\sigma_{\delta\delta}}, \tag{3.28}$$

and

$$\frac{b_{yy} - \lambda\sigma_{\varepsilon\varepsilon}}{b_{xx} - \lambda\sigma_{\delta\delta}} = \frac{(b_{xy} - \lambda\sigma_{\delta\varepsilon})^2}{(b_{xx} - \lambda\sigma_{\delta\delta})^2}$$

which is easily seen to imply

$$|\mathbf{B} - \lambda\mathbf{W}| = 0. \tag{3.29}$$

Also $\qquad U = Q_1/Q_2 = \lambda + (b_{xx} - \lambda\sigma_{\delta\delta})(\beta - \hat{\beta})^2/Q_2,$

whence, for minimum U, $\beta = \hat{\beta}$, and λ is the smaller root of (3.29). It is not difficult to show that

$$\lambda = \frac{F - \sqrt{[F^2 - 4|\mathbf{B}||\mathbf{W}|]}}{2|\mathbf{W}|} \tag{3.30}$$

where $|\mathbf{W}|$, $|\mathbf{B}|$ are the determinants of $\mathbf{W}$, $\mathbf{B}$ and

$$F = \sigma_{\delta\delta}b_{yy} + \sigma_{\varepsilon\varepsilon}b_{xx} - 2\sigma_{\delta\varepsilon}b_{xy}.$$

Equation (3.29) is familiar in canonical analysis (see e.g. Seal (1964), chapter 7), and section 6.5.

Some special cases are of particular interest. If $\sigma_{\delta\varepsilon} = 0$, it follows from (3.28) and (3.30) that

$$\hat{\beta} = \frac{b_{yy} - \phi b_{xx} + \sqrt{\{(b_{yy} - \phi b_{xx})^2 + 4\phi b_{xy}{}^2\}}}{2b_{xy}} \tag{3.31}$$

where $\phi = \sigma_{\varepsilon\varepsilon}/\sigma_{\delta\delta}$. This is essentially the same result as that given in (3.21). Thus (3.31) confirms that if δ_i and ε_i are uncorrelated only the ratio of their variances need be known for point estimation of β. In particular if $\phi = 1$, the result is independent of $\mathbf{W}$ and can be obtained from a principal component analysis based on $\mathbf{B}$, i.e. by determining the smaller latent root of

$$|\mathbf{B} - \lambda\mathbf{I}| = 0.$$

Seal (1964, chapter 6) describes principal component analysis.

When $\mathbf{W}$ is singular the result must be obtained *ab initio* and generally presents little difficulty. Thus when $\sigma_{\delta\delta} = 0$, the estimator of β is identical with the classical least squares regression estimator, a point that was made in section 3.1.

If $\mathbf{W}$ is not known it may be possible to estimate (using replication, for example, corresponding to each ξ_i, η_i) the variances and covariances in it and to use these estimates in place of the known values in the generalized least squares solution. Replication for each ξ_i, η_i implies a method of grouping, each ξ_i, η_i specifying a group. In practice groups may be based upon what are usually called *instrumental* variates. For example if (x_i, y_i) represents observations on an animal at age t_i, and more than one animal is observed at each age, each t_i may be looked upon as defining a group. In this sense W is a covariance matrix of *within* groups variability, and B represents the sums of squares and products *between* groups.

3.5. Structural relationships

These were defined in section 3.1 as exact linear relationships between random variates, and statistical problems of estimation arise when these are obscured by further random variation. Reverting to the notation of that section, we shall denote observations by (x_i', y_i') where

$$x_i' = x_i + \delta_i,$$

$$y_i' = y_i + \varepsilon_i = \beta x_i + \varepsilon_i,$$

where (x_i, y_i) are unknown realized values of (X, Y), (random) variates satisfying the relationship $Y = \beta X$. Under the assumption that the δ_i, ε_i are independent of X, Y and are independently and identically distributed for all i, having a normal distribution with zero means and variances $\sigma_{\delta\delta}$, $\sigma_{\varepsilon\varepsilon}$ we run into similar estimation problems to those arising in functional relationship when these variances are unknown. When X, Y have a singular normal distribution of unknown variance we must introduce at least one further unknown to specify our problem, namely $\mathrm{Var}(X) = \sigma_{XX}$. In an obvious notation x_i', y_i' are then realized values of Normal variates X', Y' such that

$$\mathrm{Var}(X') = \sigma_{XX} + \sigma_{\delta\delta}, \tag{3.32}$$

$$\mathrm{Var}(Y') = \beta^2 \sigma_{XX} + \sigma_{\varepsilon\varepsilon}, \tag{3.33}$$

$$\mathrm{Cov}(X', Y') = \beta \sigma_{XX}. \tag{3.34}$$

If all quantities on the right of (3.32) to (3.34) are unknown it would seem reasonable to replace the left-hand sides of these equations by their estimates based upon the observations; but this only gives us three equations for four unknowns. The parameters are said to be under-identified. Progress can be made if $\sigma_{\delta\delta}$ or $\sigma_{\varepsilon\varepsilon}$ or their ratio is known. In this respect the situation is similar to that for functional relationships; a little care is needed in the case discussed by Barnett (1967) where both $\sigma_{\delta\delta}$ and $\sigma_{\varepsilon\varepsilon}$ are known. Kendall and Stuart (1967, chapter 29) show that when $\lambda = \sigma_{\varepsilon\varepsilon}/\sigma_{\delta\delta}$ is known the result is formally the same as that given in (3.21) with x_i, y_i replaced by x_i', y_i'. They also show that if $\sigma_{\delta\delta}$ only is known

$$\hat{\beta} = \frac{\mathrm{Cov}(X', Y')}{\mathrm{Var}\, X' - \sigma_{\delta\delta}} \qquad (3.35)$$

and if $\sigma_{\varepsilon\varepsilon}$ only is known

$$\hat{\beta} = \frac{\mathrm{Var}\, Y' - \sigma_{\varepsilon\varepsilon}}{\mathrm{Cov}(X', Y')} \qquad (3.36)$$

Kiefer and Wolfowitz (1956) have given conditions for a structural relationship to be identifiable.

3.6. Types of relationship arising in practice

In this book more attention is paid to least squares regression and to functional relationships between mathematical variables than is paid to bivariate regression and to structural relationships.

It is the author's experience that the former type of models are of more practical interest. Others with experience in different fields of application may not agree, and the statement is not intended to be dogmatic. There are certain parallels in the theory for the two cases as results in this and the previous chapter have indicated.

It is of course possible that observed values of a regressor may be regarded as variates if determined in one way, but not if determined in some other way that destroys the element of randomness. We gave an example in section 1.3 when discussing heights of army recruits.

Kendall (1951, 1952) has discussed in very general terms the models in regression, structural and functional relationships.

Exercises

3.1. Verify the statement made in section 3.2 that under the assumptions made there no loss of generality results in taking α to be zero. (Lindley (1947)).

3.2. Suppose $\hat{\sigma}_\delta{}^2$ is given by (3.17). Show that under the assumption $\sum \xi_i{}^2/n$ converges to V, then $\hat{\sigma}_\delta{}^2$ converges to

$$\tfrac{1}{4}(\sigma_\delta{}^2 + \sigma_\varepsilon{}^2/\beta^2)$$

as $n \to \infty$.

3.3. Write down L^* analogous to (3.7) when δ_i, ε_i are assumed correlated with *known* covariance c. Obtain the normal equations analogous to (3.8) to (3.11) and show that (3.16), (3.17) and (3.18) still hold.

3.4. In obtaining (3.22) and (3.23) it was stipulated that $\sigma_{\varepsilon\varepsilon}\sigma_{\delta\delta} \neq \sigma_{\delta\varepsilon}^2$. Why? How might one go about the problem if $\sigma_{\varepsilon\varepsilon}\sigma_{\delta\delta} = \sigma_{\delta\varepsilon}$? (Hint: Rotate axes to give a known model for the new variates).

3.5. In (3.35) the sign in the denominator is negative. In discussing the attenuation effect in section 3.1 the corresponding sign in the expression for $E(b_1)$ was positive. Explain the reason for the difference in terms of the model under consideration and the aim in each case.

3.6. Explain why a choice of sign other than that mentioned below (3.16) would not be sensible in the problem considered there.

CHAPTER 4

Regression and Functional Relationship with Heterogeneous and Correlated Departures

4.1. Removal of departure restrictions in regression

In this section we turn again to least squares regression where a variate Y has for given x a mean $\alpha + \beta x$ and α, β are to be estimated from observations (x_i, y_i). We consider a model

$$y_i = \alpha + \beta x_i + \varepsilon_i \qquad (4.1)$$

differing from that considered in section 2.4 by dropping the requirement that the ε_i be independent from observation to observation and also the requirement that they have the same variance for all i.

Solution of the problem now invokes the principle of generalized least squares introduced by Aitken (1933, 1934). In this chapter we deal only with point estimation of β. Problems of interval estimation are dealt with in the more general multivariate and multivariable situations in chapter 5, the relevant results for the present case being easily deduced.

A typical biological situation, in which (4.1) is appropriate and ε_i and ε_j may be correlated and not necessarily of the same variance, occurs in growth studies. Suppose y_i represents the weight of an organism at time x_i and the weight of that *same* organism is measured at times $x_1, x_2, \ldots, x_n$, and a similar set of weights is obtained for a number of organisms from the same population (e.g. organisms submitted to some particular treatment) at the same times. If all the observations are plotted they will often be found to be scattered about a straight line which is often referred to as an average growth curve. It is worth commenting that in practice this can usually only

47

be regarded as a straight line over relatively short time periods in the life span of most organisms. The points associated with an individual organism will tend to show correlations between departures from the line, at least over intervals that are not very long. For example, an organism that is above average size for the population at time x_1 is likely to be above average size also at time x_2 and may even remain above average size throughout the period of study; similar arguments may be applied for those initially below average size. In this case the ε_i would tend to show a positive pattern of correlations between occasions. There are situations where organisms initially larger than average rapidly lose that advantage and end up smaller than ones that are initially below average size. In this case ε_i for small i will show a negative correlation with ε_i for large i. If we write the idealized average growth curve for a population as

$$y = \alpha + \beta x \tag{4.2}$$

our model for individuals is specified by (4.1) where the ε_i represent biological departures for an individual at time x_i from the population average weight specified by (4.2). We use the word *idealized* here to indicate a mathematical abstraction; if the population under study were one of overweight men it might be medically ideal to be well below average weight for that population!

In the context of our example, not only may successive departures be correlated for each individual, but they may well show trends such as the variance increasing or decreasing with time. The model proposed by Aitken allowed for an $n \times n$ covariance matrix $\mathbf{\Sigma}$ for $\varepsilon_1, \varepsilon_2, \ldots, \varepsilon_n$, which in practice are usually assumed to be normally distributed. In matrix notation the observational equations for an individual

$$y_1 = \alpha + \beta x_1 + \varepsilon_1,$$
$$y_2 = \alpha + \beta x_2 + \varepsilon_2,$$
$$\cdot \quad \cdot \quad \cdot \quad \cdot \quad \cdot \quad \cdot$$
$$y_n = \alpha + \beta x_n + \varepsilon_n$$

may be written

$$\mathbf{y} = \alpha \, \mathbf{1} + \beta \mathbf{x} + \mathbf{\varepsilon} \tag{4.3}$$

where $\mathbf{y}, \mathbf{1}, \mathbf{x}, \varepsilon$ are respectively column vectors of observed y values, units, observed x values and unknown ε values. Denoting the transpose of a column vector $\mathbf{a}$ by $\mathbf{a}'$, if $\mathbf{\Sigma}$ is known the likelihood function for α, β has the form

$$L = \frac{1}{(2\pi|\mathbf{\Sigma}|)^{n/2}} \exp\left\{-\frac{1}{2}(\mathbf{y}-\alpha\mathbf{1}-\beta\mathbf{x})'\mathbf{\Sigma}^{-1}(\mathbf{y}-\alpha\mathbf{1}-\beta\mathbf{x})\right\}.$$

Differentiation of $L^* = \ln L$ with respect to the scalars α, β leads to normal equations

$$\mathbf{1}'\mathbf{\Sigma}^{-1}(\mathbf{y}-\alpha\mathbf{1}-\beta\mathbf{x}) = 0, \tag{4.4}$$

and

$$\mathbf{x}'\mathbf{\Sigma}^{-1}(\mathbf{y}-\alpha\mathbf{1}-\beta\mathbf{x}) = 0, \tag{4.5}$$

which may be written

$$\mathbf{1}'\mathbf{\Sigma}^{-1}\mathbf{1}\alpha + \mathbf{1}'\mathbf{\Sigma}^{-1}\mathbf{x}\beta = \mathbf{1}'\mathbf{\Sigma}^{-1}\mathbf{y},$$

and

$$\mathbf{x}'\mathbf{\Sigma}^{-1}\mathbf{1}\alpha + \mathbf{x}'\mathbf{\Sigma}^{-1}\mathbf{x}\beta = \mathbf{x}'\mathbf{\Sigma}^{-1}\mathbf{y},$$

whence

$$\hat{\beta} = \frac{\mathbf{1}'\mathbf{\Sigma}^{-1}\mathbf{1}.\mathbf{x}'\mathbf{\Sigma}^{-1}\mathbf{y} - \mathbf{1}'\mathbf{\Sigma}^{-1}\mathbf{y}.\mathbf{x}'\mathbf{\Sigma}^{-1}\mathbf{1}}{\mathbf{1}'\mathbf{\Sigma}^{-1}\mathbf{1}.\mathbf{x}'\mathbf{\Sigma}^{-1}\mathbf{x} - \mathbf{1}'\mathbf{\Sigma}^{-1}\mathbf{x}.\mathbf{x}'\mathbf{\Sigma}^{-1}\mathbf{1}}. \tag{4.6}$$

In classical least squares regression $\mathbf{\Sigma} = \sigma^2\mathbf{I}$, whence $\mathbf{\Sigma}^{-1} = (1/\sigma^2)\mathbf{I}$. Also, since $\mathbf{x}'\mathbf{I}\mathbf{x} = \mathbf{x}'\mathbf{x} = \sum x_i^2$, $\mathbf{x}'\mathbf{y} = \sum x_i y_i$, $\mathbf{1}'\mathbf{x} = \mathbf{x}'\mathbf{1} = \sum x_i$, $\mathbf{1}'\mathbf{y} = \mathbf{y}'\mathbf{1} = \sum y_i$ and $\mathbf{1}'\mathbf{1} = n$ it follows that in this case (4.6) reduces to

$$\beta = \frac{n\sum x_i y_i - \sum y_i \sum x_i}{n\sum x_i^2 - (\sum x_i)^2},$$

which is equivalent to (2.6).

It follows immediately from (4.4) that

$$\hat{\alpha} = \frac{\mathbf{1}'\mathbf{\Sigma}^{-1}\mathbf{y}}{\mathbf{1}'\mathbf{\Sigma}^{-1}\mathbf{1}} - \beta\frac{\mathbf{1}'\mathbf{\Sigma}^{-1}\mathbf{x}}{\mathbf{1}'\mathbf{\Sigma}^{-1}\mathbf{1}} \tag{4.7}$$

which clearly reduces to

$$\hat{\alpha} = \bar{y} - \beta\bar{x}$$

when $\mathbf{\Sigma} = \sigma^2\mathbf{I}$.

Note that for the general form of $\mathbf{\Sigma}$ we cannot eliminate the constant term α by taking the origin at $(\bar{x}, \bar{y})$. The constant can only be eliminated if the origin is taken at $(\mathbf{1}'\mathbf{\Sigma}^{-1}\mathbf{x}/\mathbf{1}'\mathbf{\Sigma}^{-1}\mathbf{1}, \mathbf{1}'\mathbf{\Sigma}^{-1}\mathbf{y}/\mathbf{1}'\mathbf{\Sigma}^{-1}\mathbf{1})$.

An important special case occurs when the ε_i are uncorrelated with one another but not necessarily of equal variance; Σ^{-1} is then diagonal with its ith diagonal element the reciprocal of the variance of ε_i. We write this reciprocal $w_i = 1/\sigma_{ii}$, where $\sigma_{ii} = \text{Var}(\varepsilon_i)$. The appropriate estimators may be deduced from (4.6) and (4.7) or more directly by writing down the likelihood function for this case, which is

$$L = \frac{1}{(2\pi\sigma_{11}\sigma_{22}\ldots\sigma_{nn})^{n/2}} \exp\left\{-\frac{1}{2}\sum(y_i - \alpha_i - \beta x_i)^2/\sigma_{ii}\right\}.$$

Proceeding in the usual way the estimators for α, β are easily shown to be

$$\hat{\alpha} = \sum w_i y_i / \sum w_i - \beta \sum w_i x_i / \sum w_i \tag{4.8}$$

and

$$\hat{\beta} = \frac{\sum w_i x_i y_i - \sum w_i x_i \sum w_i y_i / \sum w_i}{\sum w_i x_i^2 - (\sum w_i x_i)^2 / \sum w_i}. \tag{4.9}$$

The w_i are called *weights*, and the estimation procedure is known as *weighted least squares*. Clearly (4.8) and (4.9) reduce to classical least squares estimators when for all i, $w_i = 1/\sigma^2 = w$, say.

In the general model Σ is seldom known, and Rao (1959) has shown that we may replace Σ by an estimate S having a Wishart distribution with f degrees of freedom. Such an estimate is available with growth measurements of the type already discussed in this section, providing measurements are made at *each* of the times x_1, $x_2, \ldots, x_n$ on *each* of p, $(p > 1)$ organisms selected from the same population. Denote by σ_{ij} the element in row i and column j of Σ (i.e. σ_{ij} is the covariance of ε_i and ε_j); if we now write y_{ik} for the observation on the kth organism at time x_i and $y_{i\cdot} = \sum_k y_{ik}/p$ for the mean of the observations on all p organisms at time x_i, an estimate s_{ij} of σ_{ij} with $p-1$ degrees of freedom is given by

$$s_{ij} = \left(\sum_k y_{ik}y_{jk} - py_{i\cdot}y_{j\cdot}\right)/(p-1).$$

The s_{ij} are the required elements of S. Elston and Grizzle (1962) have applied this method to some data on growth of the ramus bone in boys.

Rao develops appropriate formulae for interval estimation and hypothesis tests for parameters and the fitted line; these are discussed in the more general context of multiple regression in chapter 5.

4.2. Correlated departures in functional relationships

In this section, as in section 4.1, we deal only with point estimation and consider a method of generalized least squares based on Sprent (1966). We suppose n observations (x_i, y_i) to satisfy the model

$$x_i = \xi_i + \delta_i$$

$$y_i = \eta_i + \epsilon_i = \beta \xi_i + \varepsilon_i$$

where, as in chapter 3, ξ_i, η_i are unobservable values of variables ξ, η. We now drop the assumption that δ_i, ε_i are uncorrelated with δ_j, ε_j, $i \neq j$. In this section we consider the homogeneous case $\eta = \beta \xi$, but it is no longer true that $\eta = \alpha + \beta \xi$ reduces to this case if the origin is taken at the mean of the observed values; the non-homogeneous case is considered in chapter 6 where a dummy variable identically equal to 1 is associated with α, giving rise to a trivariable situation.

When the restrictions that successive departures are uncorrelated and identically distributed are dropped, it is clear from the discussion of the simpler situation in section 3.2 and 3.3 that it is going to be extremely difficult to make headway unless something is known about the variances and covariances. We shall first assume that the departures are normally distributed, and that the covariance matrix of the $2n$-variate normal distribution $(\delta_1, \varepsilon_1, \delta_2, \varepsilon_2, \ldots, \delta_n, \varepsilon_n)$ is known. All the components δ_i, ε_i are assumed, as usual, to have zero means. We denote the covariance matrix by $\mathbf{W}_0$. In chapter 3 it was assumed that $\mathbf{W}_0$ had the simple pattern

$$\mathbf{W}_0 = \begin{bmatrix} \mathbf{W} & 0 & 0 & 0 & . & . & . & 0 \\ 0 & \mathbf{W} & 0 & 0 & . & . & . & 0 \\ 0 & 0 & \mathbf{W} & 0 & . & . & . & 0 \\ . & . & . & . & . & . & . & . \\ 0 & . & . & . & 0 & \mathbf{W} & 0 & 0 \\ 0 & . & . & . & 0 & 0 & \mathbf{W} & 0 \\ 0 & . & . & . & 0 & 0 & 0 & \mathbf{W} \end{bmatrix} \tag{4.10}$$

where $\mathbf{W}$ had the form

$$\mathbf{W} = \begin{pmatrix} \sigma_{\delta\delta} & \sigma_{\delta\varepsilon} \\ \sigma_{\delta\varepsilon} & \sigma_{\varepsilon\varepsilon} \end{pmatrix}$$

and each $\mathbf{0}$ was a 2×2 zero matrix, there being n diagonal submatrices $\mathbf{W}$, one for each pair of observations. In the general case if we assume an arbitrary form for $\mathbf{W_0}$ we may partition it into 2×2 sub-matrices $\mathbf{W}_{ij}$ of the form

$$\mathbf{W}_{ij} = \begin{pmatrix} \sigma_{\delta_i \delta_j} & \sigma_{\delta_i \varepsilon_j} \\ \sigma_{\varepsilon_i \delta_j} & \sigma_{\varepsilon_i \varepsilon_i} \end{pmatrix},$$

where $\sigma_{\delta_i \delta_j}$ is the covariance between δ_i and δ_j, etc. Note that in general $\sigma_{\delta_i \varepsilon_j} \neq \sigma_{\varepsilon_i \delta_j}$ and thus the matrices $\mathbf{W}_{ij}$ are not symmetric even though $\mathbf{W_0}$ is symmetric.

The procedure used in Sprent (1966) for the simple situation discussed in section 3.4 where $\mathbf{W_0}$ had the form (4.10) required consideration of the variate

$$z_i = y_i - \beta x_i,$$

termed by Williams (1955) a null variate because it contains only residual variation. The function U given in (3.27) is essentially

$$\sum \{ z_i^2 / \mathrm{Var}(z_i) \},$$

since, in the notation of section 3.4, $\sum z_i^2 = b_{yy} - 2\beta b_{xy} + \beta^2 b_{xx}$, and $\mathrm{Var}(z_i) = \sigma_{\varepsilon\varepsilon} - 2\beta \sigma_{\varepsilon\delta} + \beta^2 \sigma_{\varepsilon\varepsilon}$.

We now describe for the general case a procedure which is based upon an intuitive generalization of (3.27). It is not clear that this is equivalent to maximum likelihood in the most general situations, although the author conjectures that it is. More work is needed on this point.

Denote the column vector of residuals for a given β by $\mathbf{z}$, i.e. $\mathbf{z}' = (z_1, z_2, \ldots, z_n)$. Let $\mathbf{\Theta}$ denote the covariance matrix of $\mathbf{z}$, then for arbitrary $\mathbf{W_0}$ a typical element of $\mathbf{\Theta}$ is

$$\theta_{ij} = \mathrm{Cov}(z_i, z_j) = \beta^2 \sigma_{\delta_i \delta_j} - \beta \sigma_{\delta_i \varepsilon_j} - \beta \sigma_{\varepsilon_i \delta_j} + \sigma_{\varepsilon_i \varepsilon_j}.$$

The procedure is to minimize

$$U = \mathbf{z}' \mathbf{\Theta}^{-1} \mathbf{z}. \tag{4.11}$$

Clearly, if for all $i \neq j$, $\theta_{ij} = 0$, and all θ_{ii} are equal, (4.11) reduces to (3.27). For given $\mathbf{\Theta}$ it is possible, at least in theory, to evaluate Θ^{-1} as a function of β and to differentiate U with respect to β to form the normal equation.

We may write

$$\mathbf{z}' = (-\beta, 1) \begin{pmatrix} \mathbf{x}' \\ \mathbf{y}' \end{pmatrix}$$

where $\mathbf{x}, \mathbf{y}$ are column vectors of the n observed variate values. Thus writing

$$\mathbf{\Phi} = \begin{pmatrix} \mathbf{x}' \\ \mathbf{y}' \end{pmatrix} \Theta^{-1} (\mathbf{x}, \mathbf{y}) = \begin{pmatrix} \phi_{xx} & \phi_{xy} \\ \phi_{xy} & \phi_{yy} \end{pmatrix}$$

it follows from (4.11) that

$$U = (-\beta \ 1) \ \mathbf{\Phi} \begin{pmatrix} -\beta \\ 1 \end{pmatrix} = \beta^2 \phi_{xx} - 2\beta \phi_{xy} + \phi_{yy}. \qquad (4.12)$$

The appearance of U in this form is deceptively simple, as the elements of $\mathbf{\Phi}$ are in general complicated functions of β.

Differentiation of (4.12) with respect to β leads to the normal equation

$$2(\beta\phi_{xx} - \phi_{xy}) + \beta^2 \frac{\mathrm{d}\phi_{xx}}{\mathrm{d}\beta} - 2\beta \frac{\mathrm{d}\phi_{xy}}{\mathrm{d}\beta} + \frac{\mathrm{d}\phi_{yy}}{\mathrm{d}\beta} = 0. \qquad (4.13)$$

If ϕ is not a function of β the solution is

$$\hat{\beta} = \phi_{xy}/\phi_{xx}. \qquad (4.14)$$

Some special cases are dealt with and a numerical example is given in Sprent (1966). In the general case iterative solutions involving repeated evaluations of Θ^{-1} are required. No investigation has been carried out of the relative merits of different methods of computation, but when $\mathbf{W_0}$ is given, an obvious method is to evaluate U for a series of values of β using a coarse spacing, and then using a finer mesh as the value of β for a minimum is tracked down. In some cases convergence may be slow. This method will generalize to the multivariable situation, while in the simple bivariable case it is possible to get an approximate value of β by graphical methods and to construct a starting mesh accordingly.

Two special cases of interest may be mentioned. The first arises if Θ is diagonal. The Θ^{-1} is also diagonal with diagonal elements $1/\theta_{ii}$ and Φ has elements

$$\phi_{xx} = \sum(x_i{}^2/\theta_{ii}), \quad \phi_{xy} = \sum(x_iy_i/\theta_{ii}), \quad \phi_{yy} = \sum(y_i{}^2/\theta_{ii})$$

whence

$$U = \sum_i \frac{\beta^2 x_i{}^2 - 2\beta x_i y_i + y_i{}^2}{\beta^2 \sigma_{\delta_i \delta_i} - 2\beta \sigma_{\delta_i \varepsilon_i} + \sigma_{\varepsilon_i \varepsilon_i}} \tag{4.15}$$

and if $\sigma_{\delta_i \delta_i} = \sigma_{\delta_i \varepsilon_i} = 0$ this reduces to the function minimized in weighted least squares regression in section 4.1.

In the second case the matrices $\mathbf{W}_{ij}$ all have the form

$$\mathbf{W}_{ij} = \begin{pmatrix} 0 & 0 \\ 0 & \sigma_{\varepsilon_i \varepsilon_j} \end{pmatrix}$$

and this implies the δ_i are all identically zero but the ε_i, ε_j are correlated; then Θ^{-1} is independent of β and (4.14) holds with

$$\beta = \frac{\mathbf{x}'\Theta^{-1}\mathbf{y}}{\mathbf{x}'\Theta^{-1}\mathbf{x}}$$

where Θ^{-1} is now the same as Σ^{-1} of section 4.1, providing we restrict the argument to the case $\alpha = 0$, a restriction made in this section but not in section 4.1.

Patterned forms of $\mathbf{W}_0$ occur in certain situations associated with designed experiments, but the sub-matrices corresponding to $\mathbf{W}_{ij}$ are then often larger than 2×2, and correspond to situations involving more than two variables. A useful paper dealing with this aspect of the subject is that by Villegas (1961).

If $\mathbf{W}_0$ is not known, but there are replicated observations corresponding to each ξ_i, η_i it is possible to proceed using an estimate of $\mathbf{W}_0$ at least as far as point estimation of β.

4.3. Regression models of the second kind

In section 4.1 we discussed a biological situation where it was reasonable to expect correlated departures from an underlying relationship. There we arrived at a model that was not very specific about the nature of such departures, requiring simply that the ε_i and ε_j associated with different x values had a known correlation matrix Σ, or at least that an estimate of Σ was available from replication.

It sometimes happens in growth studies that each individual has a characteristic growth curve of its own, the curves for the different individuals clustering together to some extent. It is then convenient to think of the parameters specifying the individual curves as having a distribution of some sort about certain mean values.

In this section we consider only the case where the curves for individuals are straight lines, each with characteristic slopes, these slopes having a probability distribution of some specified form. This model differs fundamentally from those considered earlier in that departures from the underlying curve are now associated with changes in parameters such as slope and intercept, rather than with random variates added to underlying variates or variables. This type of model has been termed a model of the second kind, and although developed largely in connection with growth studies it has wider applicability. In growth studies if we assume (as may well be true over a limited period of time) that a size measurement y_i (e.g weight or height) of the ith organism at time x_i is given by

$$y_i = \alpha_i + \beta_i x_i \qquad (4.16)$$

then we have a model of the second kind if α_i, β_i are randomly distributed about means α, β. We shall assume α_i, β_i are normally distributed about these means and the problem of interest is to estimate α and β when a set of n values (x_i, y_i) are given; at first we suppose each one is determined for a different individual.

Models of the second kind have been considered by Rao (1965a), Fisk (1967) and Nelder (1968). Nelder suggests that in addition to applications in growth studies, they may be appropriate in agricultural experiments with fertilisers. He argues that if x is the amount of fertilizer applied, and y the resulting yield of some crop, one can look upon the field plot as a black box converting input x into output y, and can argue that we know (very often accurately) the amount of fertilizer applied and the amount of crop it produces, but what we do not know are the parameters of individual black boxes, each corresponding to a different plot. If we assume they convert fertilizer to yield by a process which, for the ith plot can be represented by a model

$$y_i = \alpha_i + \beta_i x_i,$$

C

where α_i, β_i are jointly normally distributed with means α, β and covariance matrix

$$\begin{pmatrix} \sigma_{\alpha\alpha} & \sigma_{\alpha\beta} \\ \sigma_{\alpha\beta} & \sigma_{\beta\beta} \end{pmatrix}$$

we have a model of the second kind. We can use observations (x_i, y_i) on different plots to estimate α, β and, if required, their covariance matrix. We assume the model (4.16) holds exactly for individual plots. In growth studies, although (4.16) may hold approximately for individuals, there will in practice be additional random departures from the curve for an individual. We ignore this complication at the moment and deal with what we shall call a *pure* model of the second kind, where the random element is associated entirely with the parameters of the regression equations.

The treatment here is based upon that given by Nelder (1968). With the model specified it follows that y_i in (4.16) is normally distributed about $\alpha + \beta x_i$ with variance $\sigma_{\alpha\alpha} + 2\sigma_{\alpha\beta}x_i + \sigma_{\beta\beta}x_i^2$ and for n independent observations $y_1, y_2, \ldots, y_n$ the likelihood function is proportional to

$$\frac{1}{\prod_i (\sigma_{\alpha\alpha} + 2\sigma_{\alpha\beta}x_i + \sigma_{\beta\beta}x_i^2)^{1/2}} \exp -\frac{1}{2} \sum \left\{ \frac{(y_i - \alpha - \beta x_i)^2}{\sigma_{\alpha\alpha} + 2\sigma_{\alpha\beta}x_i + \sigma_{\beta\beta}x_i^2} \right\}$$

The logarithm of the likelihood, L^* is thus proportional to

$$-\sum (\ln(\sigma_{\alpha\alpha} + 2\sigma_{\alpha\beta}x_i + \sigma_{\beta\beta}x_i^2)) - \sum \frac{(y_i - \alpha - \beta x_i)^2}{\sigma_{\alpha\alpha} + 2\sigma_{\alpha\beta}x_i + \sigma_{\beta\beta}x_i^2} + k, \quad (4.17)$$

where k is a constant.

The normal equations are obtained in the usual way by differentiating with respect to α, β, $\sigma_{\alpha\alpha}$, $\sigma_{\alpha\beta}$ and $\sigma_{\beta\beta}$ and setting the derivatives equal to zero. Writing

$$w_i = \frac{1}{\sigma_{\alpha\alpha} + 2\sigma_{\alpha\beta}x_i + \sigma_{\beta\beta}x_i^2}$$

it is clear that if $\sigma_{\alpha\alpha}$, $\sigma_{\alpha\beta}$, $\sigma_{\beta\beta}$ were known, estimation of α, β would be a weighted least squares problem of the type considered in section 4.1. If these variances and covariances are not known the normal equations may be written:

$$\sum w_i(y_i - \alpha - \beta x_i) = 0, \qquad (4.18)$$

$$\sum w_i x_i(y_i - \alpha - \beta x_i) = 0, \qquad (4.19)$$

$$\sum w_i - \sum w_i^2(y_i - \alpha - \beta x_i)^2 = 0, \qquad (4.20)$$

$$\sum w_i x_i - \sum w_i^2 x_i(y_i - \alpha - \beta x_i)^2 = 0, \qquad (4.21)$$

$$\sum w_i x_i^2 - \sum w_i^2 x_i^2(y_i - \alpha - \beta x_i)^2 = 0. \qquad (4.22)$$

The above equations may be solved iteratively. Nelder points out that there is no guarantee that $\hat{\sigma}_{\alpha\alpha}$, $\hat{\sigma}_{\alpha\beta}$, $\hat{\sigma}_{\beta\beta}$ estimated from (4.18) to (4.22) will satisfy the necessary conditions for them to be elements of a covariance matrix, i.e. $\hat{\sigma}_{\alpha\alpha}$, $\hat{\sigma}_{\beta\beta}$ must both be non-negative and $\hat{\sigma}_{\alpha\beta}^2 \leqq \hat{\sigma}_{\alpha\alpha}\hat{\sigma}_{\beta\beta}$. He introduces constraints by an ingenious device to ensure that these conditions hold, introducing the following transformation of the variances:

$$\sigma_{\alpha\alpha} = \lambda \sin^2\theta,$$

$$\sigma_{\beta\beta} = \lambda \cos^2\theta,$$

$$\sigma_{\alpha\beta} = \lambda \sin\theta \cos\theta \sin\phi,$$

where $\lambda > 0$, $0 \leqq \theta \leqq \pi/2$ and $-\pi/2 \leqq \phi \leqq \pi/2$. The correlation coefficient between α, β is given by $\sin\phi$ and λ is essentially a scaling factor. If we write

$$w_i' = \frac{1}{\sin^2\theta + 2x_i \sin\theta \cos\theta \sin\phi + x_i^2 \cos^2\theta}$$

then $w_i = w_i'/\lambda$ and (4.17) may be written

$$\sum \ln(w_i'/\lambda) - \sum \frac{w_i'(y_i - \alpha - \beta x_i)^2}{\lambda} + k. \qquad (4.23)$$

Differentiating (4.23) with respect to λ and equating the derivative to zero shows that at the maximum

$$\lambda = \frac{\sum(y_i - \alpha - \beta x_i)^2 w_i'}{n},$$

whence the function to be maximized can be written

$$\sum \ln w_i' - n \ln \left[\sum w_i'(y_i - \alpha - \beta x_i)^2\right]. \qquad (4.24)$$

This has to be maximized with respect to variations in α, β, θ, ϕ. Nelder suggests starting with initial values of θ and ϕ and using these in the normal equations to obtain estimates of α and β. He then suggests using the simplex method of Nelder and Mead (1965) to minimize (4.24) with signs reversed. Details and a numerical example are given by Nelder (1968).

A striking feature of the above model is that a complicated pattern of departures from an underlying relationship $y = \alpha + \beta x$ is permitted, but we do not need a *prior* estimate of the departure covariance matrix as we did for the model considered in section 4.1. The situation becomes more complicated if we allow the model for the ith observation to take the form

$$y_i = \alpha_i + \beta_i x_i + \varepsilon_i.$$

Even if we allow the ε_i all to be independent $N(0, \sigma^2)$, one further parameter is necessarily introduced into the model.

When for given x values, $x = x_i$ we have replicated observations, as would be the case in a growth study, where measurements are made on several organisms at each of a series of times, we may use a different approach to models of the second kind. In particular, Rao (1965a) considers models of the second kind as an alternative to the model considered in section 4.1, with an estimate $\mathbf{S}$ of $\mathbf{\Sigma}$. Rao's paper may be consulted for details, and we return briefly to this method in section 5.7 where a more general case is considered. Effectively he makes use of the replication to estimate $\sigma_{\alpha\alpha}$, $\sigma_{\alpha\beta}$, $\sigma_{\beta\beta}$ and $\sigma_{\varepsilon\varepsilon}$ when appropriate.

4.4. Some special patterns of correlations

Some special patterns of correlations between departures from underlying regression or functional relationships occur in certain fields of application. This is particularly true in the study of economic time series. The relationships involved in the study of time series are seldom straight lines and can more appropriately be considered when dealing with curves. The subject is a wide one and is not dealt with in this book; this is not meant to imply that it is not important, but rather that it calls for a more specialized treatment, and such a work could more appropriately be produced by someone with more experience in this field than the author. Some of the models applicable in economic situations have been considered by Fisk (1966) and by Malinvaud (1966).

We shall, however, consider just one particular departure pattern where each departure depends upon the departure at the stage immediately before it. In the straight line context considered so far in this book it is the case where Σ as defined in section 4.1 has a form determined by a Markov Process. Suppose in n successive observations the departures take the form

$$\varepsilon_1 = \gamma_1,$$
$$\varepsilon_r = \rho\varepsilon_{r-1} + \gamma_r, \ r = 2, 3, \ldots, n$$

where γ_1 is $N(0, \sigma^2)$ and γ_r $(r > 1)$ is $N\{0, (1-\rho^2)\sigma^2\}$. With this model, where each departures depends on earlier ones only through a rather special type of linear relationship with its immediate predecessor, Σ takes the form

$$\sigma^2 \begin{bmatrix} 1 & \rho & \rho^2 & . & . & . & \rho^{n-1} \\ \rho & 1 & \rho & \rho^2 & & & \rho^{n-2} \\ . & . & . & . & . & . & . \\ . & . & . & . & \rho & 1 & \rho \\ \rho^{n-1} & . & . & . & \rho^2 & \rho & 1 \end{bmatrix}.$$

4.5. Non-linear relationships and transformation of data

So far our attention has been confined to straight-line relationships where the pattern has been blurred by the presence of random variation.

There is obviously much data to which no straight line can reasonably be fitted, despite a certain amount of wishful thinking by some experimenters. Other curves, such as polynomials, may be more appropriate, but we defer consideration of these to chapter 5.

Transformations of the variates or variables involved in a relationship may sometimes produce linearity. For example, suppose there exists an approximate relationship

$$y = \alpha e^{\beta x}, \tag{4.25}$$

then
$$\ln y = \ln \alpha + \beta x \tag{4.26}$$

and one would expect an approximate linear relationship between x and $\ln y$. If on physical, biological, economic or whatever, theory we expect (4.25) to hold approximately, a common procedure is to apply classical regression methods to $(x_i, \ln y_i)$. Sometimes methods appropriate to functional relationships are applied, but the implication of assumptions about error should not be neglected.

The classical regression model applied to (4.26) assumes that

$$\ln y_i = \ln \alpha + \beta x_i + \varepsilon_i \qquad (4.27)$$

where the ε_i are independent $N(0, \sigma^2)$, and (4.27) gives

$$y_i = \alpha e^{\beta x_i + \varepsilon_i}$$

$$= \alpha e^{\beta x_i} \varepsilon_i'$$

where $\varepsilon_i' = e^{\varepsilon_i}$.

In this case if ε_i is normally distributed, ε_i' is clearly not, and further, the error in y is multiplicative rather than additive. Whether this is a reasonable assumption in any practical situation is up to the user to decide.

It is not proposed to go into the question of transformations in detail here. The best practical advice may well be to avoid them unless there are theoretical grounds for using a particular transformation. Some evidence about validity of error assumptions may be obtained from an examination of residuals, a matter discussed in section 7.7.

Taking logarithms of both x and y is a commonly used transformation to obtain linearity. In some biological studies there is a theoretical justification for this when x, y are size measurements on different parts of an organism. A linear relationship between log x and log y is usually referred to in this context as the *simple allometry relationship*, a term introduced by Huxley (1924). The relationship may arise as follows. If x, y are the sizes of the measured organs at time t, their relative growth rates are defined as $1/x \; dx/dt$ and $1/y \; dy/dt$. For many growing organisms these are found to be roughly proportional to one another, i.e. their ratio is constant for all t so that

$$\frac{1}{y} \frac{dy}{dt} = \frac{k}{x} \frac{dx}{dt}.$$

Integration with respect to t gives

$$\ln y = k \ln x + \text{constant}.$$

Note that the base of the logarithm is immaterial in this relationship and in practice it is usual to take the base 10, making appropriate adjustments to the constants.

Application of the principle of allometry has produced interesting results, but a great deal of biological material is so variable that a straight line fitted to the original observations, their square roots, or their logarithms, will often give what appear to be equally reasonable fits although, as already pointed out in section 2.7, no direct means of comparison of these fits is available.

Transformations of a very different kind are discussed in section 9.1.

Exercises

4.1. Obtain expressions analogous to those in (2.10), (2.11), (2.12) and (2.16) for variances in weighted least squares regression where $\hat{\alpha}$, $\hat{\beta}$ are given by (4.8) and (4.9).

4.2. Show that, in general, if there are n sets of observations (x_i, y_i), equation (4.13) can be reduced to an algebraic equation of degree $12n-1$ in β. (Sprent, (1966)).

4.3. Show that (4.14) reduces to the classical least squares solution when $\sigma_{\varepsilon_i \varepsilon_i} = \sigma^2$ for all i, and all other elements of $\mathbf{W}_0$ are zero.

4.4. For the model in section 4.2 suppose each $\mathbf{W}_{ij}$ in $\mathbf{W}_0$ has the form $\mathbf{W}_{ij} = k_{ij}\mathbf{W}$ where

$$\mathbf{W} = \begin{pmatrix} \sigma_{\delta\delta} & \sigma_{\delta\varepsilon} \\ \sigma_{\varepsilon\delta} & \sigma_{\varepsilon\varepsilon} \end{pmatrix}.$$

Show that (4.11) then reduces to

$$U = \frac{\beta^2 \mathbf{x}' \mathbf{K}^{-1} \mathbf{x} - 2\beta \mathbf{x}' \mathbf{K}_\alpha^{-1} \mathbf{y} + \mathbf{y}' \mathbf{K}^{-1} \mathbf{y}}{\beta^2 \sigma_{\delta\delta} - 2\beta \sigma_{\delta\varepsilon} + \sigma_{\varepsilon\varepsilon}}$$

where $\mathbf{K}$ is a matrix with elements k_{ij}. (Villegas, (1961); Sprent, (1966).)

4.5. Investigate models of the second kind (section 4.3) when $\alpha_i \equiv 0$ for all i, i.e. $y_i = \beta_i x_i$, where β_i are independent $N(0, \sigma_{\beta\beta})$.

4.6. What restriction does the condition that Σ must be positive definition place upon the permissible values of ρ in a Markov process?

4.7. Is there a simple transformation that will reduce the problem of estimating the parameters in a relationship

$$(x-\alpha)(y-\beta) = \gamma$$

to a linear regression problem?

CHAPTER 5

Multiple Regression

5.1. Multivariate regression

In this chapter we consider the extension of various regression concepts already discussed for one regressor variate or variable to a larger number of regressors. We first consider *multivariate* normal regression. This is an extension to p, $(p > 2)$ variates of the concepts introduced in section 2.2. This type of regression is considered in detail by Anderson (1958, chapters 2 and 8).

In general we may consider the regression of any sub-set of r of the p variates on the remaining q, $q = p - r$. It will be convenient to denote the regressor variates by $X_1, X_2, \ldots, X_q$ and the dependent variates by $Y_1, Y_2, \ldots, Y_r$. Writing these as column vectors $\mathbf{X}$, $\mathbf{Y}$, we mean by the regression of $\mathbf{Y}$ on $\mathbf{X}$ the expression giving the locus of the mean value of the conditional distribution of $\mathbf{Y}$ for $\mathbf{X} = \mathbf{x}$. In matrix notation this mean is written $E(\mathbf{Y}|\mathbf{X} = \mathbf{x})$ and it is an $r \times 1$ column vector. Anderson shows that, writing $\boldsymbol{\mu}_Y$, $\boldsymbol{\mu}_X$ for the mean vectors of $\mathbf{Y}$, $\mathbf{X}$ respectively

$$E(\mathbf{Y}|\mathbf{X} = \mathbf{x}) = \boldsymbol{\mu}_Y + \boldsymbol{\Sigma}_{12}\boldsymbol{\Sigma}_{22}^{-1}(\mathbf{x} - \boldsymbol{\mu}_X) \tag{5.1}$$

where the covariance matrix $\boldsymbol{\Sigma}$ of *all* the variates written in the order $Y_1, Y_2, \ldots, Y_r, X_1, X_2, \ldots, X_q$ has been partitioned

$$\begin{pmatrix} \boldsymbol{\Sigma}_{11} & \boldsymbol{\Sigma}_{12} \\ \boldsymbol{\Sigma}_{21} & \boldsymbol{\Sigma}_{22} \end{pmatrix}$$

where $\boldsymbol{\Sigma}_{11}$ is an $r \times r$ matrix and consequently $\boldsymbol{\Sigma}_{12}$, $\boldsymbol{\Sigma}_{21}$, $\boldsymbol{\Sigma}_{22}$ are respectively $r \times q$, $q \times r$ and $q \times q$. It is assumed $\boldsymbol{\Sigma}_{22}$ is non-singular.

The vector equation (5.1) represents a set of r linear equations and is essentially a generalization of the bivariate equation (2.2).

63

This is easily seen if ρ is written in the form $\sigma_{12}/\sqrt{(\sigma_{11}\sigma_{22})}$, for then (2.2) becomes

$$E(Y|X = x) = \mu_Y - \sigma_{12}\sigma_{22}^{-1}(x - \mu_X) \qquad (5.2)$$

an analogous form to (5.1).

In practice the case of multivariate regression most commonly met is that in which $r = 1$, when (5.1) reduces to the single equation

$$y = E(Y|X_1 = x_1, X_2 = x_2, \ldots, X_q = x_q)$$
$$= \mu_Y + \boldsymbol{\Sigma}_{12}\boldsymbol{\Sigma}_{22}^{-1}(\mathbf{x} - \boldsymbol{\mu}_X)$$

and this equation is more familiar in one of the forms (5.3) or (5.4), i.e.

$$y = \mu_Y + \beta_1(x_1 - \mu_1) + \beta_2(x_2 - \mu_2) + \ldots + \beta_q(x_q - \mu_q) \qquad (5.3)$$

or
$$y = \beta_0 + \beta_1 x_1 + \beta_2 x_2 + \ldots + \beta_q x_q, \qquad (5.4)$$

where $\mu_1, \mu_2, \ldots, \mu_q$ are the elements of $\boldsymbol{\mu}_X$ and $\beta_1, \beta_2, \ldots, \beta_q$ are the elements of $\boldsymbol{\Sigma}_{12}\boldsymbol{\Sigma}_{22}^{-1}$ and

$$\beta_0 = \mu_Y - \boldsymbol{\Sigma}_{12}\boldsymbol{\Sigma}_{22}^{-1}\boldsymbol{\mu}_X$$

Equations (5.3) and (5.4) are alternative forms of the multiple regression equation for a single variate Y on $X_1, X_2, \ldots, X_q$; this is a special case of the regression of a vector variate $\mathbf{Y}$ on a vector variate $\mathbf{X}$.

Either of the regression equations (5.1) or (5.3) may be estimated from sets of observations of the variates providing n, the number of sets, is sufficiently large to give a non-singular estimate of $\boldsymbol{\Sigma}$ whenever $\boldsymbol{\Sigma}$ is non-singular. The sample means based on the observations provide the estimates of $\boldsymbol{\mu}_X$, $\boldsymbol{\mu}_Y$. If $\mathbf{S}$, the estimator of $\boldsymbol{\Sigma}$, is partioned in a similar manner, the coefficients β_i, $i = 1, 2, \ldots q$ are estimated by the elements of $\mathbf{S}_{12}\mathbf{S}_{22}^{-1}$ in (5.3) and (5.4) and β_0 in the latter is estimated by

$$\hat{\beta}_0 = \bar{y} - \mathbf{S}_{12}\mathbf{S}_{22}^{-1}\bar{\mathbf{x}}$$

where $\bar{y}$, $\bar{\mathbf{x}}$ are means of the observations.

Equations (5.3) or (5.4) are equations to q dimensional spaces or hyperplanes in a space of $q+1 = p$ dimensions. The system of equations (5.1) consists of r independent linear equations when $\boldsymbol{\Sigma}$ is non-singular and defines a sub-space of $p-r = q$ dimensions.

If there are p variates altogether, in the case $r = 1$ there are p possible choices of a dependent variate, and thus p regressions of one variate on the remainder in all. For the case $r = 2$ with two dependent variates, these may be selected in $_pC_2$ ways, giving rise to $p(p-1)/2$ pairs of regression equations, and in general when $r = s$ there are $_pC_s$ sets of s linear regression equations. It is of course possible to obtain regression equations involving only a sub-set of the given variates; deletion of variates effectively reduces the dimension of the space under consideration.

In referring to the equations (5.1), (5.3) and (5.4) as linear, it must be stressed that the term is used in its mathematical sense and essentially implies linearity in the *coefficients*; In p dimensions (5.1) only defines a straight line if Σ_{22} is a one dimensional matrix (i.e. a scalar). This corresponds to the case of one regressor variate and $p-1$ dependent variates, and (5.1), known then as a vector equation to a straight line, contains $p-1$ independent equations. By independent equations we mean that none of them are linear combinations of any of the remainder. Other aspects of relationship between **X** and **Y** are dealt with in section 6.5.

5.2. Multivariable regression

We now consider the regression of a variate Y on (mathematical) variables $x_1, x_2, \ldots, x_q$. We consider in this section the case where y is normally distributed about a mean which is a linear function $\beta_0 + \boldsymbol{\beta}'\mathbf{x}$ of the q variables with variance (which may not be known) σ^2, independent of **x**.

The basic problem of interest is estimation of $\boldsymbol{\beta}$ given n sets of observed values $(x_{1i}, x_{2i}, \ldots x_{qi}, y_i)$, or more briefly $(\mathbf{x}_i, y_i)$. The problem is analogous to the least squares regression problem of section 2.4. It will simplify notation somewhat if we associate the constant β_0 in the equation for this case analogous to (5.4) with a dummy variable x_0 that always takes the value one. This introduces one additional variable. It is also easily demonstrated that we may absorb the constant term β_0 without altering the values of $\boldsymbol{\beta}$ by taking the origin at the mean of the observations, i.e. at a point $(\bar{\mathbf{x}}, \bar{y})$. We shall suppose for the moment that the origin is at the mean of the observations. We use matrix notation. Let **y** represent a column vector of n observed values of y, i.e. the transpose of **y** is $\mathbf{y}' = (y_1, y_2, \ldots, y_n)$. Corresponding to y_i we have the values x_{1i},

$x_{2i}, \ldots, x_{qi}$ of the regressors. We may form an $n \times q$ matrix, $\mathbf{X}$ of these observed regressor variable values, when the observational equations take the form

$$\mathbf{y} = \mathbf{X}\boldsymbol{\beta} + \boldsymbol{\varepsilon},$$

$\boldsymbol{\varepsilon}$ being an $n \times 1$ vector of departures, the elements of which are independently and identically distributed $N(0, \sigma^2)$, where σ^2 is usually unknown. As in section 2.4 maximum likelihood estimation is easily shown to be equivalent to minimization of

$$E = (\mathbf{y} - \mathbf{X}\boldsymbol{\beta})'(\mathbf{y} - \mathbf{X}\boldsymbol{\beta}),$$

the matrix equivalent of (2.9). The normal equations are obtained by differentiation with respect to each element of $\boldsymbol{\beta}$ and equating the derivatives to zero in the usual way and are found to be

$$(\mathbf{X}'\mathbf{X})\boldsymbol{\beta} = \mathbf{X}'\mathbf{y}$$

with solutions

$$\hat{\boldsymbol{\beta}} = (\mathbf{X}'\mathbf{X})^{-1}\mathbf{X}'\mathbf{y} \qquad (5.5)$$

With modern computing facilities inversion of the $q \times q$ matrix $\mathbf{X}'\mathbf{X}$ is not very time-consuming for values of q as high even as 20. Although regression equations have been formulated for more than 20 regressor variables their interpretation is usually so difficult as to make them virtually meaningless. It is easily verified that $\hat{\beta}$ given by (5.5) has the same value as would be obtained using $\mathbf{S}_{12}\mathbf{S}_{22}^{-1}$ in the notation introduced in the multivariate situation of section 5.1. This equivalence enables us to use the above estimation procedure even when we have a mixture of normal regressor variates and regressor variables in the one problem. As in chapter 2 we shall simply speak of regressors when they may be either normal variates or variables.

The term *multiple regression* is often used in reference to both multivariable and multivariate regressions. In practice one usually has little worry in classical multiple regression situations about whether one is dealing with variates or variables or a mixture, providing they truly measure what they purport to measure (i.e. are error free).

When the origin has been transferred to the mean it is easily shown on transforming back to the original variates that β_0 is estimated by

$$\hat{\beta}_0 = \bar{y} - \hat{\beta}_1 \bar{x}_1 - \hat{\beta}_2 \bar{x}_2 - \ldots - \hat{\beta}_q \bar{x}_q.$$

In practice the terms in $\mathbf{X'X}$ are computed as the sums of squares and products of deviations about the mean rather than transforming the original data by change of origin. For example, writing $\mathbf{S} = \mathbf{X'X}$ the element s_{uv} is computed as

$$s_{uv} = \sum_i x_{ui} x_{vi} - \left(\sum_i x_{ui}\right)\left(\sum_i x_{vi}\right)/n.$$

The formulae obtained in section 2.6 for standard errors and confidence limits extend to this multiple regression case. We note that the residual sum of squares is

$$(\mathbf{y} - \mathbf{X}\hat{\boldsymbol{\beta}})'(\mathbf{y} - \mathbf{X}\hat{\boldsymbol{\beta}}) = \mathbf{y'y} - \hat{\boldsymbol{\beta}}'\mathbf{X'X}\hat{\boldsymbol{\beta}} = \mathbf{y'y} - \hat{\boldsymbol{\beta}}\mathbf{X'y},$$

and this can be used to obtain an estimate of σ^2 with $n-q$ degrees of freedom. The total sum of squares is $\mathbf{y'y}$ with n degrees of freedom, and $\hat{\boldsymbol{\beta}}\mathbf{X'y}$ represents the regression sum of squares with q degrees of freedom (one for each parameter that has been estimated). In the usual analysis of variance the sum of squares about the mean is referred to as the *total* sum of squares and it has $n-1$ degrees of freedom. The lost degree of freedom corresponds to estimation of one parameter, effectively β_0. The covariance matrix of $\boldsymbol{\beta}$ is easily seen to be $\sigma^2(\mathbf{X'X})^{-1}$ and standard errors and confidence limits for individual β_i may be obtained in a manner similar to that indicated for one regressor in section 2.6.

The regression coefficients $\beta_1, \beta_2, \ldots, \beta_q$ are often referred to as partial regression coefficients. The meaning of this is perhaps more easily seen if we consider two regressor variables. For convenience we take the origin at the mean. The regression of y on x_1 alone is estimated by

$$y = \hat{\beta}x_1, \tag{5.6}$$

where

$$\hat{\beta} = \left(\sum_i x_{1i} y_i\right)/\left(\sum x_{1i}\right)^2. \tag{5.7}$$

The regression of y on x_1 *and* x_2 is estimated by

$$y = \hat{\beta}_1 x_1 + \hat{\beta}_2 x_2, \tag{5.8}$$

and solution of (5.5) for $\hat{\beta}_1$ gives

$$\hat{\beta}_1 = \frac{\sum_i x_{2i}^2 \sum x_{1i} y_i - \sum_i x_{1i} x_{2i} \sum x_{2i} y_i}{\sum x_{1i}^2 \sum x_{2i}^2 - \left(\sum x_{1i} x_{2i}\right)^2}. \tag{5.9}$$

From (5.7) and (5.9) it is clear that in general $\hat{\beta} \neq \hat{\beta}_1$. Further, when fitting x_1 alone the regression sum of squares is $\hat{\beta}\sum x_{1i} y_i$ and when fitting x_1 *and* x_2 it is $\hat{\beta}_1 \sum x_{1i} y_i + \hat{\beta}_2 \sum x_{2i} y_i$, and it follows again from (5.7) and (5.9) that $\hat{\beta}_1 \sum x_{1i} y_i \neq \hat{\beta} \sum x_{1i} y_i$. In fitting x_1 and x_2 the two terms in the regression sum of squares appear at first sight to be each associated with one variable, but this is not so in general since $\hat{\beta}_1$ depends upon x_2 as well as x_1 values, and $\hat{\beta}_2$ similarly depends upon both regressors. In section 5.3 we discuss orthogonal variates and here the single terms in the sum of squares, equation (5.14), are each associated with one variate, but this is a very special case.

In the general case, (5.8) represents geometrically in the three dimensional space of y, x_1, x_2, the plane of best fit in the least squares sense—i.e. it minimizes the sum of squares of departures in the y direction; but (5.6) represents the least square line of best fit to the projections of the observed points on the (x_1, y)—plane. Clearly the direction cosines of the normal to this line will in general specify a different direction to the direction cosines of the normal to the plane.

So far we have assumed that $\mathbf{X'X}$ (and equivalently $\mathbf{S}_{22}$ in multivariate regression) is non-singular, so that $\mathbf{X'X^{-1}}$ exists and there is no difficulty in solving the normal equations uniquely. An important practical situation in which $\mathbf{X'X}$ is singular arises in experimental design and analysis theory, and is taken up again in chapter 7. At this stage we illustrate one way in which singularity may arise and discuss implications concerning the nature of partial regression coefficients in this case. Suppose we wish to estimate the regression of a variate y on two variables x_1 and x_2 using the data in table 5.1.

TABLE 5.1. Observed values of a variate y and variables x_1 and x_2.

x_1	0	1	2	3	4	5
x_2	-3	-1	1	3	5	7
y	-11	-6	-2	3	8	13

Writing the regression equation

$$y = \beta_0 x_0 + \beta_1 x_1 + \beta_2 x_2 \qquad (5.10)$$

where $x_0 = 1$, the matrix $\mathbf{X'X}$ is easily computed as

$$\begin{pmatrix} 6 & 15 & 12 \\ 15 & 55 & 65 \\ 12 & 65 & 94 \end{pmatrix}$$

This matrix has zero determinant, and is thus singular, so there is no unique solution (5.5). It is not difficult to see that the trouble arises because there is a linear relation between x_1 and x_2 for the given data, namely

$$x_2 = 2x_1 - 3.$$

Thus if we know *either* x_1 *or* x_2 the other is uniquely determined, and if we estimate the regression of Y on x_1 a knowledge of x_2 is of no additional use in estimating the mean value of Y given x_1, for this mean is given by (5.10) which can be written

$$y = \beta_0 + \beta_1 x_1 + \beta_2 (2x_1 - 3)$$

$$= \beta_0' + \beta_1' x_1, \text{ say,}$$

where $\qquad\qquad \beta_0' = \beta_0 - 3\beta_2 \qquad\qquad (5.11)$

and $\qquad\qquad \beta_1' = \beta_1 + 2\beta_2. \qquad\qquad (5.12)$

Now estimators of β_0', β_1' can be uniquely determined by considering the regression of Y on x_1, and it follows from (5.11) and (5.12) that in this singular case we may choose any one of β_0, β_1 or β_2 arbitrarily, and when we do so the other two are automatically fixed. This is a direct consequence of singularity that reflects itself in a degree of arbitrariness in the estimators. It illustrates in a limiting case what is implied by calling β_1 and β_2 partial regression coefficients. From (5.12) we see that if β_2 is set equal to zero, the regression of Y on x_1 gives all the information on the conditional mean of Y obtainable from the data, and if β_1 is set equal to zero, the regression of Y on x_2 gives the same information, i.e. if either x_1 or x_2 appears in the regression equation the other is irrelevant; if both are included in the regression equation, the contribution each makes to the regression sum of squares may be arbitrarily

divided between the two subject to the constraints $\beta_0\sum x_0y+\beta_1\sum x_1y$
$+\beta_2\sum x_2y = \beta_0'\sum x_0y+\beta_1'\sum x_1y = \beta_0''\sum x_0y+\beta_2''\sum x_2y$ where the summations are over the corresponding products for the data and β_0'', β_2'' are the estimated regression coefficients of Y on x_2, the other coefficient estimators corresponding to those in (5.11) and (5.12).

When regression equations are being estimated from observational data it is seldom that x_1 and x_2, or any other pair of observed regressors, will follow a perfect linear relationship as in the above example. However, in many practical situations there is an approximate linear relationship between regressors. In these circumstances there may be little gain in measuring both and if each were highly correlated with Y in a prediction problem one would normally take as regressor the cheaper or easier regressor to measure. By a prediction problem we mean one in which the aim is to predict the value of Y for observed values of variables $x_1, x_2, \ldots, x_q$ using a regression equation established from a set of observations $(x_{i1}, x_{i2}, \ldots, x_{iq}, y_i)$.

A situation where a prediction equation is of use and there is a choice between regressor variables highly correlated with one another occurs in some experimental situations in plant physiology.

Plant physiologists are often interested in the *dry weight* of plants. Determination of this is clearly a destructive measurement, but physiologists are often interested in the dry weight increment over a period and would therefore like to know the dry weight of particular plants both at the beginning and end of that period. There is no difficulty in getting the dry weight at the end of the period, but it must be estimated at the beginning from non-destructive measurements. A technique used to do this is to take a random sample of plants at the beginning of the period and measure not only their dry weight but also other characteristics that do not require destruction for measurement; physiologists may also make use of records taken earlier in the life of the plant that could be expected to influence dry weight. These non-destructive records are also taken on the plants not destroyed at the beginning of the period. Using the initial sample a regression equation may be set up for the regression of dry weight upon the other measurements and this can be used to predict the dry weight of plants not destroyed at this time. General questions of choice of regressors are discussed in section 5.5, but we illustrate the point about correlated regressors

70

by considering dry weight estimations of this type for young apple trees. In an experiment on these at East Malling Research Station the author had made available to him records of weight at planting, the subsequent extension growth and leaf area, number of leaves and some other records at the time the trees were pulled up for dry weight determinations. In this experiment weight at planting and final leaf area were found to be very suitable predictor regressors. However, leaf area was a time-consuming measurement to take on the plants that were not to be grubbed and also carried with it a risk of damaging the leaves. Leaf number is a much easier variable to record. Therefore, bearing in mind that leaves vary greatly in size, it is of some interest to know whether leaf number is as good, or nearly as good, a predictor of plant dry weight as leaf area. In this particular experiment it was found that it was, leaf area and leaf number being in general highly correlated with one another and with plant dry weight.

For good prediction it is necessary to account for as much of the variability as possible by regression. However, this does not mean that when a regression equation is set up as many regressors as possible should be introduced, for it is clear that if we have n sets of observations of Y and the regressors and there are $n-1$ regressor variables there will be no degrees of freedom remaining for error and an exact fit will be obtained for the given data. This, of course, does not mean that an additional value of Y will be exactly predicted for a further set of regressor values.

The experimenter who measures everything possible in the belief that by doing so he will obtain a more useful linear regression equation for prediction must remember this pitfall. In addition, there is often no very sound reason for assuming a *linear* regression model will be at all appropriate for a large number of regressors.

We saw as a consequence of (5.11) and (5.12) that if x_1 and x_2 are perfectly linearly related, one may be omitted from the regression equation altogether. If, as in the case of leaf area and leaf number in the experiment on apple trees, leaf area (x_1) is highly correlated with leaf number (x_2), both being highly correlated with dry weight (y), then it will be found that there is little difference between the sums of squares

$$\beta_0 \sum x_{0i} y_i + \beta_1 \sum x_{1i} y_i + \beta_2 \sum x_{2i} y_i$$

71

where $x_0 \equiv 1$, obtained by fitting both x_1 and x_2, and the sums of squares

$$\beta_0' \sum x_0 y_i + \beta_1' \sum x_{1i} y_i \text{ or } \beta_0'' \sum x_0 y_i + \beta_2'' \sum x_{2i} y_i,$$

resulting from fitting only x_1 or only x_2. Note, however, that β_1 and β_1' may be numerically very different from one another, as also may be β_2 and β_2''.

The topic of partial regression is closely related to that of partial correlation. Good accounts of the relationship are given by Weatherburn (1946) and by Anderson (1958).

5.3. Orthogonal regressors

In section 5.2 we noted that in general the coefficient of x_1 when we consider the regression of Y on x_1 and x_2 is·not the same as the coefficient when we consider the regression of Y on x_1 alone. We considered the extreme case of a linear relationship between x_1 and x_2, which implies perfect correlation between them. Another extreme case occurs when x_1 and x_2 are completely uncorrelated. This implies that $\sum x_{1i} x_{2i} = 0$. More generally, if any two regressors x_u and x_v are uncorrelated this implies $\sum_i x_{ui} x_{vi} = 0$. Further, if x_1, x_2, ..., x_q are all uncorrelated this implies that $\mathbf{X'X}$ is a diagonal matrix and thus so is $(\mathbf{X'X})^{-1}$. Thus (5.5) reduces to a set of q equations

$$\beta_t = \left(\sum_i x_{ti}^2\right)^{-1}\left(\sum x_{ti} y_{ti}\right)$$

$$t = 1, 2, \ldots, q.$$

(5.13)

Thus the general problem reduces from that of solving q simultaneous equations in q unknowns to the much simpler one of solving q equations, each in one unknown. Also, if an additional regressor x_{q+1} is now to be included in the equation this does not require the recomputation of $\beta_1, \beta_2, \ldots, \beta_q$ if x_{q+1} is not correlated with earlier regressors.

We call regressors having this property orthogonal variates or variables. The difficulty is that observed regressors seldom have this property. Theoretically it is possible to change our regressors $x_1, x_2, \ldots, x_q$ to a new set $\xi_1, \xi_2, \ldots, \xi_q$ which are linear functions of the x_i and which are orthogonal to one another. The regression of Y on $\xi_1, \xi_2, \ldots, \xi_q$ can then be computed, and if the regression is required in terms of the x_i this can be obtained by expressing each ξ_j as the appropriate linear function of the x_i. Any term may be omitted from the regression in terms of the ξ_i without recomputing all the other coefficients. The difficulty is that in the general case it requires as much computational labour to determine the ξ_j as it does to fit the regression equation in terms of the x_i; also, in general, each ξ_j will involve some or all of the x_i.

Wishart and Metakides (1953) in a paper that deserves to be better known, consider a method which involves essentially the fitting first of x_1, then the fitting of a regressor orthogonal to x_1 that is expressible in terms of x_1 and x_2 only, then a regressor orthogonal to those already fitted that involves x_1, x_2 and x_3 only, and so on. They include an example, although their computational procedure is not particularly well suited to an electronic computer.

If $\hat{\gamma}_r$ is the estimated regression coefficient of ξ_r in terms of orthogonal regressors, then the regression sum of squares is

$$R_{SS} = \sum_r \left(\hat{\gamma}_r \sum_i \xi_{ri} y_i \right) \qquad (5.14)$$

in an obvious notation. If a regressor ξ_s is dropped from the equation, R_{SS} is reduced by an amount $\hat{\gamma}_s \sum_i \xi_{si} y_i$, and each term within the brackets on the right of (5.14) gives the contribution of one orthogonal regressor to the regression sum of squares independently of the others.

In the example already given for apple trees, if x_1 represented leaf number and x_2 leaf area, and the Wishart and Metakides method were used with x_1 as the first regressor, the second regressor orthogonal to it would be a linear function of x_1 and x_2 and could be expected to make only a small contribution to the regression sum of squares relative to that made by x_1 alone. Note that the total sum of squares attributable to fitting of the orthogonal regressors ξ_1, ξ_2 must equal the sum of squares due to fitting x_1 and x_2.

With the exception of the case to be considered in detail in section 5.4 orthogonal regressors are not very often used in practice because the saving in computation at the fitting stage is lost, due to the difficulty of determining appropriate orthogonal regressors as functions of the x_i.

Orthogonality principles are, however, the basis of several methods of adding or deleting a regressor from a regression equation

73

in such a way as to avoid complete matrix inversion at each stage. These methods depend upon the use of adjustment formulae for the elements of the matrix $(\mathbf{X}'\mathbf{X})^{-1}$ corresponding to one more or one fewer regressors. The classic paper on this topic is that by Cochran (1938). Problems associated with the logic, as distinct from the mechanics, of adding or deleting a regressor are taken up again in section 5.5.

5.4. Polynomials and other special linear regression models

In sections 5.1 and 5.2 we considered situations where the regressors were not in general independent (this is another reason for objecting to the term independent variates or variables for regressors). There is in fact no reason why x_2 should not equal x_1^2 or log x_1 or $x_1^{3/5}$.

A case of particular interest arises when we have

$$x_1 = x, \, x_2 = x^2, \, x_3 = x^3, \ldots, x_q = x^q.$$

If we also write $x_0 = 1 \, (= x^0)$ the regression equation in the form (5.4) becomes

$$y = \beta_0 + \beta_1 x + \beta_2 x^2 + \ldots + \beta_q x^q. \qquad (5.15)$$

The methods of this chapter then allow a polynomial to be fitted to paired observations (x_i, y_i). In practice it is seldom necessary or advisable to fit polynomials of degree exceeding 3 or 4. A difficulty that often arises if the methods of section 5.2 are applied directly concerns round-off errors in matrix inversion and estimation of coefficients. In many practical cases the coefficients become progressively smaller as the power of x increases, and while it may suffice to give β_0 and β_1 to one or two decimal places, β_2 and β_3 will be of a different order of magnitude and must be given to a certain number of significant figures if the contribution of the corresponding terms in (5.15) are to be correctly estimated. For fitting a quadratic, Healy (1963) has suggested a procedure to reduce round off difficulties.

A well known practical and labour-saving method based on the principle of orthogonality is readily available if the x values are equally spaced. In this case appropriate $\xi_1, \xi_2, \ldots, \xi_q$ in section 5.4 become *orthogonal polynomials*, ξ_r being a polynomial of degree r. These have been tabulated in several places. Fisher and Yates (1963), for example, tabulate the values of ξ_r and other subsidiary details needed for computation of sums of squares, standard errors, etc., for values of n (the number of equally spaced x values) from $n = 3$ to $n = 75$. For $n > 5$ polynomials up to ξ_5 only are given, and this is ample for most practical purposes. In the introduction they give comprehensive instructions for their use, so we do not give details here. Theoretically there is no reason why orthogonal polynomials should not be fitted to unequally spaced points, but tables are not then available and the computational difficulties mentioned in section 5.3 for orthogonal regression in general also apply.

Certain other relationships can also be brought into the linear regression model, bearing in mind that *linear* here implies linear in the coefficients. For example, given a set of n values of (Y, x_1, x_2) it is possible to fit a regression equation of the form

$$y = \beta_0 + \beta_1 x_1 + \beta_2 x_1^2 + \beta_3 x_1 x_2,$$

by introducing new regressors

$$x_1' = x_1, \, x_2' = x_1^2, \, x_3' = x_1 x_2.$$

It is not, however, possible to fit a relationship of the form

$$y = \beta_0 + \beta_1 (x_1 - \beta_2)^{-1}$$

by linear regression methods. This is essentially a non-linear regression problem of the type discussed in chapter 10.

5.5. Choice of regressors

The advent of electronic computers has reduced the tedium of multiple regression computations appreciably, and there are few large machines whose software does not include at least one program for multiple regression computations. In section 5.2 we mentioned the possibility of a large number of potential regressors being available. Sometimes to handle all these may exceed the capacity of the machine handling the data. The author once read a report of an experiment in which the only available machine could handle less than half of the regressors available, and so the experimenters solemnly suggested trying as many combinations selected at random of t regressors as time would permit, t being the number of regressors the machine could handle. For the machine they were using, t equalled 20; putting aside the important practical question of how one interprets a linear regression equation with 20 regressors after one has obtained it, the random selection of regressors from a larger batch is an example of using statistical procedures to the exclusion of common sense. In any large batch of regressors (and in the particular study there were a great number of meteorological variables) some of the regressors will be highly correlated with one another as well as with the dependent variate. From the remarks in section 5.2 it should be clear that there may be little gain in including two such highly correlated regressors if they are both about equally well correlated with Y. A word of caution is necessary here as the argument in section 5.2 was framed in simple terms of two competing regressors, and the pattern of partial correlations between more than two regressors is also important in influencing the choice of regressors.

It is seldom that, with a large number of possible regressors, one is not in a position to know that some will almost certainly be more important than others on commonsense reasoning rather than on purely statistical grounds. One may also suspect that certain regressors will tell much the same story and are likely to be alternatives to one another. This was the case with leaf area and leaf number in our example on the young apple trees. For other regressors there may be extreme doubt whether they are worth including in the equations. It may then be worth trying each as an additional regressor and studying its effect on the regression sum of squares. Some caution is needed in determining objective criteria for adding or deleting a regressor in these circumstances. It has already been indicated that if any physical interpretation is to be put on a regression equation, this is going to be extremely difficult with more than a few regressors. If the aim of an experiment is primarily to provide a prediction equation rather more regressors may be permissible, bearing in mind the remarks on this point in section 5.2.

The advent of computers readily able to form regression equations with any or all of from one to as many as 30 or 40 regressors has led to attempts to devise criteria for adding or deleting regressors from equations on the basis of whether they do or do not contribute *significantly* to the regression sum of squares. This intuitively reasonable procedure has the serious practical limitation that it does not lead to unique results. Depending on the initial choice of regressors, different equations containing different sub-sets of the set of all possible regressors can be obtained. For example, if one sought the five best regressors to include in a prediction equation, one would usually get a different result if one built up the equation one regressor at a time to the result obtained if one started with an equation containing the complete set of all possible regressors and eliminated them until only five were left.

At the time of writing there is no universally accepted way of dealing with the problem of selecting regressors. In most practical problems one or two regressors should probably be included almost automatically because they are easy to obtain and known to be highly correlated with the variate to be predicted. For example, if heights of children are being predicted from a series of records that include age, it seems almost inconceivable that any better or simpler observation could be included; one would feel suspicious of any regression equation that left it out on the grounds that it could well be unnecessarily complicated and that an easy-to-get observation was being replaced by some measurement of a type likely to be more costly and difficult to obtain.

Lindley (1968) has discussed the problem of choice of regressors in multiple regression in the framework of decision theory, taking a Bayesian approach. In this context he argues that the solution to the problem depends upon the use that is to be made of the regression equation and considers two cases. The first is its familiar use in prediction, and the second is its use to set Y at some fixed value by suitable control of the regressor values. Using his approach he finds the two problems have different solutions.

His study is of considerable theoretical interest both to members of the Bayesian school and to non-Bayesians, but it is not very clear how useful his methods will be in practice. His arguments depend upon carefully set out assumptions and the reader should refer to the paper for details. Essentially, the decision to include or exclude a regressor should, in Lindley's opinion, be decided by minimizing a loss function which depends not only upon accuracy of any prediction that is made or control that is achieved, but on the cost of basing the regression on the chosen sub-set of regressors.

In the simple case of a prediction problem with the regressors independent and the cost c_i of observing each x_i being additive, his analysis gives the very reasonable result that x_i should be observed if and only if

$$\{E(\hat{\beta}_i)\}^2 \, \mathrm{Var}(x_i) > c_i. \tag{5.16}$$

In this rather special case the result implies that a regressor is worth observing if its variance (in the case of a mathematical variable this implies approximately the spread of its values) is large enough, or its regression coefficient is large enough in modulus, both uses of the word *large* implying large in comparison with the cost c_i of making an observation.

It is worth noting that (5.16) implies that selection does not depend upon the accuracy with which β_i is estimated, i.e. it does not depend upon the standard error of the estimate. Lindley shows that this fact is still true for the prediction problem when the above assumptions about independence of regressors and additivity of cost no longer hold, providing his other assumptions about the nature of the decision procedure and the independence of the distribution of Y on the data actually observed are obeyed. This last assumption would seem to be a key one in Lindley's analysis.

In the control problem, even with x_i assumed independent and costs taken to be additive, Lindley shows that x_i should be included in the control equation if and only if

$$E(\hat{\beta}_i{}^2) \, \mathrm{Var}(x_i) > c_i. \tag{5.17}$$

Comparing (5.17) with (5.16) we see that the decision whether or not a regressor is to be included in the control problem depends upon the precision with which the corresponding regression coefficient is estimated.

Lindley considers application of his methods to sets of data where there is a choice of four possible regressors. With this small number of regressors it is possible to work out the residuals and costs for fitting all combinations of one, two, three or four regressors. This is a simple task on a computer and the output is not too large with so few regressors. With a large number of regressors both computer time and the volume of output may prove embarrassing. Systematic approaches to the computational problems of selecting sub-sets of regressors and fitting regression equations in a way that avoids unnecessary effort have recently been discussed by Garside (1965), Beale, Kendall and Mann (1967), Newton and Spurrell (1967) and Hocking and Leslie (1967), among others.

5.6. Multiple regressions with correlated departures

When ε_i and ε_j are correlated we are in a similar position to that in section 4.1, but we now have more than one regressor. Again we require some knowledge of the pattern of departure correlations before we can proceed even to point estimation. If their covariance matrix Σ is known in advance, the method of generalized least squares extends immediately to this case. Note that Σ here is the same as in section 4.1 and not that in section 5.1. If Σ is not known, but can be estimated from replication the procedure introduced by Rao (1959) may be used. This is probably the most common practical situation, but we present the theory first for known Σ. The observational equations may be written

$$\mathbf{y} = \mathbf{X}\beta + \varepsilon$$

where, as in section 5.2, $\mathbf{y}$ is an $n \times 1$ vector of observed values of the dependent variate, $\mathbf{X}$ is an $n \times q$ vector of observed regressor values, β is a $q \times 1$ vector of coefficients and ε is an $n \times 1$ vector of departures. We suppose the n elements of ε to be normally distributed with zero means and known covariance matrix Σ.

The likelihood function for the observations is then

$$L = \frac{1}{(2\pi|\Sigma|)^{n/2}} \exp\left\{ -\frac{1}{2}(\mathbf{y}-\mathbf{X}\beta)'\Sigma^{-1}(\mathbf{y}-\mathbf{X}\beta) \right\}. \qquad (5.18)$$

Taking logarithms it immediately follows that the estimators $\hat{\beta}$ are the values of β that minimize

$$(\mathbf{y}-\mathbf{X}\beta)'\Sigma^{-1}(\mathbf{y}-\mathbf{X}\beta).$$

Differentiating, and equating the derivatives to zero gives the normal equations

$$\mathbf{X}'\Sigma^{-1}\mathbf{X}\beta = \mathbf{X}'\Sigma^{-1}\mathbf{y}$$

with solutions

$$\hat{\beta} = (\mathbf{X}'\Sigma^{-1}\mathbf{X})^{-1}\mathbf{X}'\Sigma^{-1}\mathbf{y}. \qquad (5.19)$$

If Σ is not known, but an estimate of it is available having a Wishart distribution with f degrees of freedom, Rao shows that Σ may be replaced by this estimate, S, in (5.19). Such estimates are available in situations similar to that described for growth studies in section 4.1 and more generally where, for a fixed set of x values, there is more than one value of Y for each x, and the total number of observations is sufficient to provide a non-singular estimate of Σ.

In section 4.1 the idea of fitting an average growth curve was briefly considered, this curve taking the form

$$y = \alpha + \beta x,$$

where x is a measure of time. When observations at different times were made on the same organisms we indicated that Rao's (1959) method may be appropriate. Under a model assuming correlated departures we may in a more general situation use (5.19) with S in place of Σ to fit a polynomial, writing $x_i = x^i$, $i = 0, 1, 2, \ldots, q$. Rao gives an example of a quadratic fitted to growth data.

Even for equally spaced points orthogonality properties do not carry over directly to this model, for clearly $X'\Sigma^{-1}X$, whose inverse appears in (5.19), will only be diagonal if the variables are such that $x_r'\Sigma^{-1}x_s = 0$, $r \neq s$, where x_r, x_s are column vectors of the rth and sth regressors. With estimated Σ, Σ^{-1} is replaced by S^{-1} in the orthogonality conditions. Although it is theoretically possible to transform the regressors to a set that possesses such an orthogonality property the procedure is of limited practical usefulness.

No tabulation corresponding to that for equally spaced points in classical least squares polynomial fitting is possible as such polynomials would have to satisfy a relationship of the form $\xi_r'\Sigma^{-1}\xi_s = 0$, and this means that ξ_r, ξ_s depend upon Σ (or S in the case of estimation of the covariance matrix), a matrix that is different for each particular problem.

Sprent (1965) has shown that it is possible to use tabulated orthogonal polynomials to obtain the generalized least squares fit with Σ estimated. The method is, however, of limited usefulness as it requires the tabulation of all orthogonal polynomials up to degree $n-1$. A covariance adjustment is made to coefficients of all ordinary orthogonal polynomials of degree less than or equal to q, based upon their correlation in this more general model with higher order coefficients. Tables giving all orthogonal polynomial values for equally spaced points up to degree $n-1$ for $n \leq 26$ have been given by De Lury (1950). The procedure is only likely to be worthwhile if the polynomial is to be fitted to observations at a small number of x values, i.e. if n is small, or if there is doubt about the degree of the polynomial that will be required to give an adequate fit. Note that the correlation between ordinary orthogonal polynomials comes about in this case because they satisfy the relationship $\boldsymbol{\xi}_r'\boldsymbol{\xi}_s = 0$, and for them $\boldsymbol{\xi}_r'\boldsymbol{\Sigma}^{-1}\boldsymbol{\xi}_s \neq 0$, so they do not give rise to a diagonal matrix in the normal equations based on (5.19).

Finally in this section we give, without proof, some formulae obtained by Rao (1959) for testing adequacy of fit and for confidence limits when $\boldsymbol{\Sigma}$ is estimated by $\mathbf{S}$ having a Wishart distribution with f degrees of freedom. We give them in a form appropriate when r values of Y are observed for each value of x. These formulae, applicable to polynomial regression, are particular forms of more general results given by Rao, his results being applicable to the general multiple regression problem with correlated departures. Our notation differs slightly from that given by Rao, but to conform with his conventions we shall use $\mathbf{S}$ now to denote the dispersion matrix per observation as estimated from data. If the regression is fitted to the mean of the Y observations at each x value and $\mathbf{e} = \mathbf{y} - \mathbf{X}\hat{\boldsymbol{\beta}}$ is the vector of residuals, Rao gives a test for adequacy of fit of a polynomial of degree q. It is based on the statistic

$$T = \frac{r\mathbf{e}'\mathbf{S}^{-1}\mathbf{e}}{f}. \tag{5.20}$$

He shows that if a polynomial of degree q provides an adequate fit, then

$$(f+2-n+q)T/(n-q-1)$$

has an F-distribution with $n-q-1$ and $f+2-n+q$ degrees of freedom.

A little consideration of the form of (5.20) shows that it generalizes the concept of testing deviations from regression against an independent estimate of error in an analysis of variance after the manner demonstrated for a very simple case in table 2.3.

To obtain confidence limits for the parameters $\hat{\beta}_i$ when a polynomial of degree q is deemed to provide a satisfactory fit, the diagonal elements of $(\mathbf{X}'\mathbf{S}^{-1}\mathbf{X})^{-1}$, the covariance matrix of the estimators, must be evaluated. The off-diagonal elements of this matrix may also be required if one is interested in the estimators of estimable functions of the parameters or confidence limits for the whole curve. These questions are discussed by Rao together with a numerical example. As an example of one such formula we quote his result for $\alpha\%$ confidence limits for β_i, namely

$$\hat{\beta}_i \pm t \sqrt{\left\{\frac{c_{ii}f(1+T)}{r(f-n+q+1)}\right\}} \tag{5.21}$$

where c_{ii} is the ith diagonal element of $(\mathbf{X'S^{-1}X})^{-1}$ and t is the upper $100-\alpha$ percentile of $|t|$ with $f-n+q+1$ degrees of freedom.

In most practical cases, if the parameters are estimated by classical least squares, ignoring the correlations in departures, the point estimates will not be very different to those obtained when the correlations are taken into account, but confidence limits on the other hand depend very much upon the form of $\mathbf{\Sigma}$. In many practical cases the correct limits, taking account of correlations between the departures, are wider than those computed by classical least squares ignoring correlations; these latter are of course incorrect when correlations are present. For the simple case of a straight line (a polynomial of degree one), Elston and Grizzle (1962) give an example that shows very clearly the influence of assumptions about correlations upon confidence limits. In section 9.3 we take up the question of bias due to ignoring correlations.

5.7. Models of the second kind

These models have already been considered in section 4.3 for the case of one regressor. The concept easily extends to q regressors when we consider each experimental unit (e.g. animal in a biological investigation, town or other economic unit in an investigation by an economist, machine in an industrial process) to have a regression equation of its own, a characteristic of that equation being that for a particular equation the coefficients are random variates. We shall assume these coefficients to be normally distributed about some mean. For the kth unit we may write the equation corresponding to an observation $(x_{1i}, x_{2i}, \ldots x_{qi}, y_i)$ on that unit as

$$y_i = \beta_{0k} + \beta_{1k}x_{1i} + \beta_{2k}x_{2i} + \ldots + \beta_{qk}x_{qi} + \varepsilon_{ki} \tag{5.22}$$

where $\qquad \beta_{rk} = \beta_r + \beta'_{rk} \qquad r = 0, 1, 2, \ldots q,$

where β_r is the mean value of the variate having the value β_{rk} for the kth unit. In practice β_{rk} and β_{sk} will usually be correlated. In many practical cases it is sufficient to assume the ε_{ki} are for all k, i independently distributed $N(0, \sigma^2)$. Sometimes the ε_{ki} are taken to be identically zero.

When nothing is known about the variances and covariances of the β_{rk}, Nelder's method described in section 4.3 for the estimation of β_r and their variances and covariances, can be extended to q regressors. However, if there is replication of measurements on the same organisms at successive times of the type that enables the covariance matrix of departure from a mean curve to be calculated in the manner described in section 5.6, Rao (1965a) has shown that we may use this replication to estimate the covariance matrix of the β_r and at the same time test whether a model of the form (5.22) fits our data more satisfactorily than a model of the type used in section 5.6. For the particular case of fitting a polynomial of specified degree, q, say, Rao gives details. If a polynomial of degree q of the form (5.22) can be fitted to each individual, and the ε_{ki} are assumed independent $N(0, \sigma^2)$, then Σ of section 5.6 will take the form

$$\Sigma = \begin{pmatrix} \Sigma_R & 0 \\ 0 & \sigma^2 I \end{pmatrix} \qquad (5.23)$$

where Σ_R is a $(q+1) \times (q+1)$ matrix of variances and covariances of regression coefficients (including β_0) and I is the unit matrix of order $n-q-1$. If S is partitioned in like manner

$$S = \begin{pmatrix} S_{11} & S_{12} \\ S_{21} & S_{22} \end{pmatrix}$$

Rao provides a two-part test to decide whether S is compatible with the form (5.23) for Σ. The first part compares $|S|$ with $|S_{11}||S_{22}|$ and this is essentially a test whether the population equivalents of S_{12}, S_{21} can reasonably be zero matrices. The second statistic tests whether S_{22} is compatible with a population equivalent $\sigma^2 I$. A further example of the use of these procedures has been given by Pearson and Sprent (1968), who concluded that their particular data on hearing loss associated with age was more compatible with the model in section 5.6 than with a model of the second kind.

Exercises

5.1. Investigate more fully, paying particular attention to the phrase in italics, the statement in section 5.1 that 'The multiple regression equations (5.1) and (5.3) may be estimated from sets of observations of the variates providing n, the number of sets, is *sufficiently large to give a non-singular estimate* of Σ whenever Σ is non-singular.'

5.2. Verify that for classical least squares multiple regression there is no loss of generality in taking the origin at the mean of the observations.

5.3. Verify that for classical least squares regression the estimated covariance matrix of $\hat{\beta}$ is $\hat{\sigma}^2(X'X)^{-1}$. What is the estimated standard error of the estimate of the mean of y corresponding to a given value x_k of x?

5.4. Fit polynomials of degree 2 and of degree 3 to the following data (i) by direct computation of the coefficients of the powers of x, and (ii) by using tables of orthogonal polynomials.

x	0	1	2	3	4	5
y	4	4	6	8	11	12

5.5 Show that if we fit polynomials using the ordinary tables of orthogonal polynomials to data where the departures are correlated, a covariance adjustment based upon the correlations of these coefficients with those of higher order polynomials is required.

(Sprent, 1965).

5.6. Derive the test statistic given in section 5.6 for adequacy of fit of a polynomial of degree q when the errors are correlated. Verify the expression for confidence limits given in (5.21). Examine also the question of confidence limits for the whole curve. (Rao, 1959).

5.7. Discuss regression models of the second kind for multiple regression analogous to those given for a single regressor in section 4.3. (Nelder, 1968).

5.8. Derive formally the tests outlined in section 5.7 for determining whether replicated data accords with a model of the second kind with the departures from individual curves, ε_{ki} assumed independent $N(0, \sigma^2)$. (Rao, 1965a).

CHAPTER 6

Multidimensional Functional Relationships and Canonical Analysis

6.1. Multiplicity of functional relationships

In this chapter we extend the notions of linear functional relationships developed in chapter 3 to p variables and consider the problem of estimating such relationships between mathematical variables from observations with an additive random element, or departure, associated with some or all of the variables. For notational convenience we alter that of chapter 3 and refer to the observed variates as

$$x_1, x_2, \ldots, x_p.$$

We suppose that we have n sets of observations of these variates, and we denote the ith set by

$$x_{1i}, x_{2i}, \ldots, x_{pi},$$

where $\quad x_{ki} = \xi_{ki} + \varepsilon_{ki}, k = 1, 2, \ldots, p; i = 1, 2, \ldots, n, \qquad (6.1)$

where the *unknown* ξ_{ki} are supposed to satisfy *at least one* relationship of the form

$$\sum_{j=1}^{p} \beta_j \xi_j = 0 \qquad (6.2)$$

or in an obvious vector notation

$$\boldsymbol{\beta}' \boldsymbol{\xi} = 0.$$

For uniqueness it is necessary to apply a constraint to the β_j, and the ones most commonly used in practice are to put one of the β_j equal to $+1$ or -1, e.g. to write the equation

$$\xi_p = \beta_1\xi_1 + \beta_2\xi_2 + \ldots + \beta_{p-1}\xi_{p-1}. \tag{6.3}$$

Another constraint sometimes used is to put $\sum\beta_j^2 = 1$. Note that all the variables ξ_j are on an equal footing and there is no division corresponding to that of dependent variates and regressor variates or variables. In particular, writing the functional relationship in the form (6.3) implies no special property for ξ_p. The relationship may be perfectly validly rewritten

$$\xi_1 = \beta_2'\xi_2 + \beta_3'\xi_3 + \ldots + \beta_p'\xi_p,$$

and this will represent the same relationship as (6.3) if $\beta_2' = -\beta_2/\beta_1$, etc.

A symmetric form of the relationship has been specified in (6.2). The non-homogeneous form

$$\sum_j \beta_j\xi_j + \beta_0 = 0,$$

can always be reduced to this form by introduction of a dummy variable $\xi_0 \equiv 1$.

The phrase *at least one* used in reference to (6.2) is important, for in practice it will often be found that data will satisfy more than one independent linear relationship at least approximately. Failure to appreciate this simple possibility has perhaps caused more confusion in this field than any other single factor. It is essential to appreciate the geometric meaning both of single relationships and two or more together. In p dimensions a single linear equation represents a hyperplane or space of $p-1$ dimensions. The coefficients of the ξ_j in such a relationship are proportional to the direction cosines of the normal to that hyperplane, and if the constraint is $\sum\beta_j^2 = 1$, they are in fact the direction cosines.

If we have two independent linear relationships, each represents a space of $p-1$ dimensions. Together they represent the *intersection* of these two spaces which is a space of $p-2$ dimensions. Similarly, three independent linear relationships represent a space of $p-3$ dimensions and so on. Finally, $p-1$ such relationships represent a space of one dimension, i.e. a straight line in p-dimensional space. For example, in three dimensions it is well known that an equation of the form

$$a_1x_1+a_2x_2+a_3x_3 = 0$$

is the equation of a plane through the origin. Similarly

$$b_1x_1+b_2x_2+b_3x_3 = 0,$$

represents the equation to another plane through the origin. If they are not identical their intersection is a straight line.

An important point is that a straight line defined as the intersection of two planes is unique, but an *infinity* of pairs of intersecting planes may all define the same line, for *any* two planes intersecting on that line define it.

In two dimensions when the points used for determination of the line give a fairly good fit, linear regression methods usually give a similar estimate of the slope of a line to that based on more sophisticated functional relationship methods, since in either case a supposedly unique line is being determined. Any differences arise from different hypotheses about how observed points depart from the line.

On the other hand, with more than two dimensions we may be concerned with a number of non-unique relationships that hold approximately for the observed variates. For example, in three dimensions, if all points lie approximately on a straight line, arbitrarily choosing one variate as the dependent variate y and getting the regression on the other two will define a plane; one may reasonably hope that the required line will lie in, or allowing for departures, very nearly in, that plane, but a further plane is required to determine the precise direction of the line. It is intuitively reasonable that we should attempt where possible to define lines by planes orthogonal to one another, as errors in determining the planes due to random departures will tend then to have a minimum effect on the positioning of the line of intersection. The idea may be extended to more than three dimensions, where a sub-space of lower dimension should be determined if possible as the intersection of orthogonal higher dimensional spaces. In the case of correlated departures a modified principle of orthogonality will be required.

In determining functional relationships we want techniques that will tell us if we are making the right assumptions about the number

87

of linear relationships, as well as estimating them for us. We may be interested in certain individual relationships, as well as in the total space they define. In chapter 8 we shall see how some of these determinations work in practice.

6.2. Determination of a single linear functional relationship

When there is only one linear functional relationship the methods of chapter 3 extend immediately to p variables. Equation (3.27) is replaced by an equation that may conveniently be written in matrix notation as

$$U = \boldsymbol{\beta}'\mathbf{B}\boldsymbol{\beta}/\boldsymbol{\beta}'\mathbf{W}\boldsymbol{\beta} \tag{6.4}$$

where $\boldsymbol{\beta}$ is the $p \times 1$ vector of coefficients in the relationships, subject to a suitable constraint; $\mathbf{B}$ is the matrix of sums of squares and products between observed points with elements

$$b_{kl} = \sum_{i=1}^{n} x_{ki}x_{li}$$

and $\mathbf{W}$ is the $p \times p$ covariance matrix of the ε_{ki}, assumed the same for all i. Differentiation of (6.4) assuming the constraint $\boldsymbol{\beta}'\mathbf{W}\boldsymbol{\beta} = 1$, leads to a solution where the coefficients are the elements of the latent vector corresponding to the smallest latent root of

$$|\mathbf{B} - \lambda\mathbf{W}| = 0. \tag{6.5}$$

We next consider extension of the methods of section 4.2 to p variates, restricting attention at first to the case where there is only one underlying relationship between the variables associated with the observed variates. Now the ε_{ki} corresponding to different points may be correlated with one another and $\mathbf{W}_0$ of section 4.2 is replaced by a $pn \times pn$ matrix and the $\mathbf{W}_{ij}$ become $p \times p$ matrices; $\boldsymbol{\Theta}$ becomes an $n \times n$ natrix, and θ_{ij} becomes the quadratic form $\boldsymbol{\beta}'\mathbf{W}_{ij}\boldsymbol{\beta}$; $\boldsymbol{\Phi}$ becomes a $p \times p$ matrix $\mathbf{X}'\boldsymbol{\Theta}^{-1}\mathbf{X}$ where $\mathbf{X}$ is the $n \times p$ matrix of observed variate values and U, the function to be minimized, becomes

$$U = \boldsymbol{\beta}'\boldsymbol{\Phi}\boldsymbol{\beta}. \tag{6.6}$$

Taking partial derivatives with respect to $\boldsymbol{\beta}$ with the arbitrary constraint $\beta_p = -1$ leads to a set of $p-1$ normal equations

$$2\boldsymbol{\beta}'_m\boldsymbol{\phi}_i + \boldsymbol{\beta}'_m\frac{\partial\boldsymbol{\Phi}}{\partial\beta_i}\boldsymbol{\beta}_m = 0, \tag{6.7}$$

where $$\boldsymbol{\beta}'_m = (\beta_1, \beta_2, \ldots, \beta_{p-1}, -1),$$

and $\boldsymbol{\phi}_i$ is a column vector of elements of the ith row of $\boldsymbol{\Phi}$, ($i = 1$, $2, \ldots, p$) and $\partial\boldsymbol{\Phi}/\partial\beta_i$ is the matrix of derivatives of elements of $\boldsymbol{\Phi}$ with respect to β_i.

Special cases considered in section 4.2 have their analogues for p variates. When the elements of $\mathbf{W}_0$ have to be estimated from replication, for example, estimates can be used in place of the true known $\mathbf{W}_0$.

6.3. Confidence limits for a single relationship

In the bivariate situation in chapters 3 and 4 we postponed consideration of interval estimates of the parameters. The situation is not very satisfactory in general, and more work requires to be done on this point. We shall consider only the case of homogeneous errors where equation (3.27) or its generalization to p dimensions, (6.4), represents the function to be minimized.

A simple approach is the following. The numerators of (3.27) or (6.4) are sums of squared residuals each with variances given by the denominators. If n is the number of sets of observations, U will have a χ_n^2 distribution under the hypothesis that a linear relationship exists.

Consider the two-dimensional case where (3.27) is appropriate. For a chosen probability level α we may determine from tables of χ_n^2 a value K such that $P(\chi_n^2 > K) = \alpha$. Thus, if there is a linear relationship between ξ and η it follows that with probability $1 - \alpha$

$$U = \frac{\sum_i (y_i - \beta x_i)^2}{\beta^2 \sigma_{\delta\delta} - 2\beta\sigma_{\delta\varepsilon} + \sigma_{\varepsilon\varepsilon}} \leqq K. \tag{6.8}$$

If the minimum of U is greater than K it is reasonable to accept this as evidence against the hypothesis of a linear relationship, i.e. the value is *significant* at the α probability level. For all values of β for which (6.8) holds, the hypothesis of a linear relationship would be acceptable; thus acceptable values of β must satisfy the inequality

$$\beta^2 b_{xx} - 2\beta b_{xy} + b_{yy} \leqq K(\beta^2 \sigma_{\delta\delta} - 2\beta\sigma_{\varepsilon\delta} + \sigma_{\varepsilon\varepsilon}) \tag{6.9}$$

If the equality is taken in (6.9) we obtain a quadratic in β. The roots of this equation give confidence limits for β at the $1 - \alpha$ probability level. If they are unreal this implies that there is no value of β

for which the hypothesis of a linear relationship would prove acceptable. If the minimum of U is considerably less than K there may be a wide range of acceptable β values. A rather anomalous situation is created by the fact that if we choose $\alpha = 0 \cdot 05$ say, we may find the minimum of U is greater than the corresponding K, and thus reject the hypothesis of linearity at the $0 \cdot 05$ level. At the same time we may find that if we take $\alpha = 0 \cdot 01$ the increased value of K may result in the minimum of U being now below the new K value, so that we accept the possibility of a straight line. Attention was drawn to this anomaly by Healy (1966), who pointed out that the above approach led to the conclusion that either there was no line, or if there was one, it was extremely well determined. That this is indeed the interpretation is clear from figure 3.1. If a line exists, it must cut or pass close to all the ellipses if they represent reasonable confidence regions for the true points. If the confidence ellipses are enlarged by increasing the confidence level from, say, 95 to 99 per cent, the chances of finding a line to cut all the ellipses is increased. The so-called confidence limits associated with the equality in (6.9) are not confidence limits in the usual sense, but define sets of acceptable values of β when we accept at a certain level the hypothesis of linearity. As the numerator in U is a measure of the distance of the fitted line from the observed points, it follows that if this is small, relative to the diameters of the confidence ellipses for (ξ_i, η_i), there may be a large range of acceptable values of β.

Brown (1957) considered the generalization of the above line of argument to p variates and in place of the quadratic obtained by taking the equality in (6.9) he obtained conics or hyperconics in the parameter space that defined joint confidence regions for the parameters. He also showed that the envelope to the family of possible relationships was a conic. Details are given by Kendall and Stuart (1967).

If the error covariance matrix is estimated with f degrees of freedom, the above arguments using a χ^2 distribution can be adapted by determining K from an F distribution with n and f degrees of freedom.

Beale (1966) suggested that in many cases it may be more appropriate to consider the distribution of $U - U_{\min}$, which he conjectured would have approximately a χ_p^2 distribution. While this is reasonable intuitively as a generalization of the concept of a standardized sum of squares for a linear regression, no detailed study appears to have been made of its properties for functional relationships. The use of functions of this type in non-linear regression is described by Beale (1960).

Creasy (1956) considers in detail confidence limits and hypothesis tests about β for the case $\sigma_{\varepsilon\delta} = 0$, and $\lambda = \sigma_{\varepsilon\varepsilon}/\sigma_{\delta\delta}$ is known (see exercise 6.4). Her results are also discussed by Madansky (1959) and Lindley and El-Sayyad (1968).

6.4. Determination of more than one linear relationship

With p variates we have already pointed out that there may be as many as $p-1$ independent linear relationships. With errors independent from point to point and having the same distribution at each point with covariance matrix $\mathbf{W}$, these linear relationships may be estimated with coefficients determined by the latent vectors corresponding to certain small latent roots of (6.5), viz.

$$|\mathbf{B} - \lambda\mathbf{W}| = 0.$$

It is reasonable to associate a linear relationship with each latent root that is sufficiently small. We discuss other latent roots of this equation in section 6.5. An approximate relationship occurs for each root λ_i that is sufficiently small for us to regard its departure from zero as attributable to ε_{ki}. The elements of the latent vector corresponding to such a λ_i give the estimated coefficient values in a linear relationship. They are proportional to the direction cosines of the normal to the hyperplane.

When $\mathbf{W}$ is estimated, by replication, say, and there are n_i observations corresponding to a true point $\xi_{1i}, \xi_{2i}, \ldots, \xi_{pi}$ on the hyperplane, and we wish to test the hypothesis that there are $p-k$ independent relationships, but only have estimates x_{ki} of the ξ_{ki}, an appropriate test has been given by Bartlett (1947). It assumes a normal distribution for the departures ε_{ki}. The roots of (6.5) are written in descending order of magnitude as $\lambda_1, \lambda_2, \ldots, \lambda_p$ and if $N = \sum n_i$, the appropriate test statistics is

$$T_p = \left(N - 1 + \frac{p+n}{2}\right) \ln \prod_{j=k+1}^{p} \left\{1 + \frac{(n-1)\lambda_j}{N-n}\right\}. \qquad (6.10)$$

Under the hypothesis that $p-k$ relationships exist, T_p has a χ^2

distribution with $(p-k)(n-k-1)$ degrees of freedom. In practice the procedure is to perform the test with $k = p-1$ (i.e. test for one linear relationship), and then repeat the test with $k = p-2$, proceeding in this manner until a significant result is first obtained. Practical procedures for obtaining latent roots and vectors for (6.5) have been described by Seal (1964, chapter 7). Some practical considerations in applying the theory of linear functional relationships under various assumptions are given in chapter 8.

6.5. Canonical analysis

In sections 6.2 and 6.4 we were concerned with the small latent roots and vectors of the determinantal equation (6.5). This equation is well known as the fundamental equation in the subject of canonical analysis. In general it will have m non-zero latent roots, where m is the lesser of $n-1$ and p. The latent vectors corresponding to the non-zero latent roots of (6.5) give the coefficients of what are termed canonical variates. For the case $n = 2$ we may look upon $\mathbf{B}$ as a multiple of the covariance matrix between the two groups and $\mathbf{W}$ as a measure of the variation within each group. In this special case it is well known that the one non-zero latent root determines the vector of coefficients in the linear function that best discriminates between the two groups. It specifies the direction of a variate which is a linear function of the original variates and has the property that its between groups variance is a maximum relative to that within groups. In practice it is usual to standardize this variate so that its within groups variance is unity, i.e. denoting the variate by

$$z = \mathbf{cx}$$

where $\mathbf{x}$ is the vector of observational variables and $\mathbf{c}$ is the latent vector corresponding to the non-zero root of (6.5), $\mathbf{c}$ is so scaled that $\mathbf{c'Bc}$ is a maximum subject to $\mathbf{c'Wc} = 1$.

When $n > 2$ there will in general be more than one non-zero latent root, and the corresponding latent vectors, associated with the latent roots in descending order of magnitude, denoted by

$$\mathbf{c}_1, \mathbf{c}_2, \ldots, \mathbf{c}_m$$

will have the property that $\mathbf{c}_i'\mathbf{Bc}_i$ is maximized subject to $\mathbf{c}_i'\mathbf{Wc}_i = 1$, $\mathbf{c}_i'\mathbf{Wc}_j = 0, j < i$.

Any non-zero latent roots that, using the test statistic (6.10), do not differ significantly from zero may, as already indicated, imply zero population roots. The corresponding latent vectors determine what might be called an *error space* is the sense that between groups variation is not significantly greater than that within groups in the directions so specified. The corresponding canonical variates are often called null variates. They provide our estimated linear functional relationships in the sense that they are uncorrelated with group differences.

The canonical variates derived from the larger latent roots and vectors are correlated with differences between groups.

The concept of canonical correlations is often applied to situations where variates are of two types. In section 5.1 we considered the partitioning of p normal variates into two sets, one consisting of r variates $Y_1, Y_2, \ldots, Y_r$ and the other of $q = p - r$ variates $X_1, X_2, \ldots, X_q$. In section 5.1 we discussed the regression of **Y** on **X**. We are sometimes interested in determining linear functions in each set of variates as highly correlated as possible with one another. This is achieved by selecting first the linear functions of $Y_1, Y_2, \ldots, Y_r$ and $X_1, X_2, \ldots, X_q$ that exhibit maximum correlation. We speak of these linear functions as the first canonical variates for the correlation of **Y** on **X**. The second canonical variates are a pair uncorrelated with the first but, subject to this condition, showing as high as possible a correlation with one another, and so on. It can be shown that the coefficients in the **Y** variates are given by the latent vectors corresponding to the latent roots in descending order of

$$|\mathbf{\Sigma}_{12}\mathbf{\Sigma}_{22}^{-1}\mathbf{\Sigma}_{21} - \lambda\mathbf{\Sigma}_{11}| = 0, \tag{6.11}$$

and those for the **X** variates by the corresponding latent vectors associated with the roots of

$$|\mathbf{\Sigma}_{21}\mathbf{\Sigma}_{11}^{-1}\mathbf{\Sigma}_{12} - \lambda\mathbf{\Sigma}_{22}| = 0, \tag{6.12}$$

where $\mathbf{\Sigma}$ is partitioned as in section 5.1. In practice $\mathbf{\Sigma}$ is usually estimated from the data.

The above is the barest outline of the concepts of canonical variates and canonical correlations, intended only to indicate their relationship to the main theme of this book. Examples of the application of these ideas will be found in section 8.1.

Readers interested in details of the theory and practical use of canonical variates and canonical correlations should consult a text such as Seal (1964, chapter 7) or Anderson (1958, chapter 12). An extensive literature on the subject stems from fundamental work by Hotelling (1936). Papers by Bartlett (1947) and Williams (1955, 1967) are especially relevant to the relationship of canonical analysis to the topics in this book.

Component analysis and factor analysis are other techniques in the general field of multivariate analysis that are being increasingly used in analysis of data involving many variates. We refer the reader to Seal's book for further details.

Exercises

6.1. Verify the validity of the extension of the expression (3.27) when $p = 2$ to the general form (6.4) when $p \geqq 2$.

6.2. Obtain p variate analogues of the results given in (4.14) and (4.15) in section 4.2.

6.3. Obtain the p variate analogue of the result given in exercise 4.4.
(Villegas, (1961)).

6.4. For a bivariate functional relationship show that if $\sigma_{\delta\delta} = \sigma_{\varepsilon\varepsilon}$ and $\sigma_{\delta\varepsilon} = 0$, then under the null hypothesis that $\beta = 0$,

$$t = \{(n-2)r^2/(1-r^2)\}^{1/2},$$

where $r = b_{xy}/\sqrt{(b_{xx}b_{yy})}$, has a t distribution with $n-2$ degrees of freedom.   (Creasy, (1956); Kendall and Stuart (1967, chapter 29)).

6.5. Show that the rth canonical correlation (i.e. the correlation coefficient between the rth canonical variates) between $\mathbf{Y}$ and $\mathbf{X}$, the normal variates considered in section 6.5, is equal to the rth largest root of the determinantal equation

$$\begin{vmatrix} -\lambda\mathbf{\Sigma}_{11} & \mathbf{\Sigma}_{12} \\ \mathbf{\Sigma}_{21} & -\lambda\mathbf{\Sigma}_{22} \end{vmatrix} = 0.$$

(Anderson, (1958, chapter 12)).

CHAPTER 7

Some Applications of the Classical Regression Model

7.1. The use of regression analysis

In earlier chapters we discussed various models for the determination of lines, planes or hyperplanes in the presence of random departures in our observations from these spaces. Suggestions about applications, although a few were made, have been incidental to the main theme. We now consider various ways in which regression equations obtained by classical least squares may be examined and used in statistical analysis of data. Some of the techniques extend in a fairly obvious way to situations with heterogeneous or correlated errors, but these extensions are not discussed in this chapter. Techniques more directly concerned with functional and structural relationships are dealt with in chapter 8.

Among the problems we consider are the comparison of lines, including tests for parallelism and identity, concurrence, abrupt changes of slope. We also deal with rather special regression models associated with experimental design. In section 7.2 a rather special type of prediction problem is discussed. Some miscellaneous topics are also mentioned, including the testing of validity of the model by an examination of residuals. The choice of topics reflects to some extent the personal interests of the author, but it is hoped they will at least give some indication of problems arising in regression analysis and methods of tackling them.

A useful discussion of many general aspects of regression has been given by Cox (1968).

We have stressed in sections 2.2, 2.3 and elsewhere that in bivariate regression problems we may consider either the regression of X on Y or of Y on X and if both variates appear at random in our data we can use that data to estimate *either* regression. We pointed out as early as section 1.3 that if the experimenter deliberately selected the values of one of his variates, X, then his observations are no longer random variate values but are best regarded as observational or mathematical variables. This is the case, for example, in specifying a growth curve where x represents a time of measurement and the measurements are made at regular intervals, e.g. weekly, monthly, annually, etc.

In the bivariate case where there is a choice of regressor and dependent variates the choice of the explanatory variate as the regressor seems logical when we are considering a cause and effect relationship; e.g. for Galton's investigation discussed in sections 1.4 and 2.3, mid-parent height is clearly the explanatory variate. In bivariate prediction problems it is logical to take the variate to be predicted as the dependent variate. We remark in passing that an important property of any predicting equation is its stability. When using it there is an implicit assumption that conditions do not change as far as the model is concerned. Numerous examples can be quoted of a change in circumstances which will make a relationship worthless for prediction. An example often quoted from elementary physics is the relationship between pressure and volume of a gas at constant temperature. If we determine this relationship at a *fixed* temperature, we can use it to predict pressure given volume or volume given pressure *only* as long as the temperature is maintained at that fixed temperature. The behaviour as temperature changes is immediately evident to anyone with a knowledge of Boyle's Law and Charles's Law. These laws are themselves capable of further refinement to allow for properties of real gases as opposed to conceptual perfect gases.

In section 5.2 we discussed the estimation of the dry weight of a plant from certain non-destructive measurements. The regression equation was based upon a random sample from a larger population and while applicable to other members of that population it may not, and indeed without confirmatory evidence would not, be permissible to use the same equation on plants of a different variety or species, or on plants of a different age, or even on the same plants at different ages. We can only do this if we have shown the law to be of wide general validity, i.e. that the same regression equation held independently of these other factors. Invariance of a prediction equation is an important and desirable property and tests for identity of the underlying population regression equations corresponding to different sets of data are therefore important, and will be discussed in section 7.4.

Standard errors for estimates of Y for a given x present no difficulty whether x is a variate or a variable value. Appropriate formulae were given in section 2.6 for classical least squares with one regressor.

Some difficulty arises when the regressor is a variable and we want to predict the value of x that has given rise to an observed value, y, of Y. We cannot solve this problem by obtaining the regression of x on Y, for the regression of a variable on a variate is a meaningless concept. We discuss this problem in detail in the next section.

7.2. Inverse estimation

In section 2.5 we discussed a calibration procedure that might be used by a chemist. In that example we considered an observation y obtained by a quick or cheap method which was related to a true value x that the chemist wanted to determine. We discussed how he could obtain his regression equation by observing Y for various fixed x (a variable) and pointed out that in future he would want to use that equation to estimate x, given an observed value of Y, and indicated that this is not a simple statistical problem. We now consider this question; to simplify the algebra slightly we transfer our origin to $(\bar{x}, \bar{y})$, the mean of the observations used in obtaining the regression equation. Writing $x' = x - \bar{x}, y' = y - \bar{y}$ the regression equation can be written

$$E(Y') = \hat{\beta}x'.$$

It follows that an observed value of Y', say y'_q will arise from an *unknown* value of x', say x'_q such that

$$y'_q = \hat{\beta}x'_q + e,$$

where e is an unknown residual representing departure from the estimated regression equation. Thus

$$x'_q = \frac{y'_q - e}{\hat{\beta}},$$

and since e has expectation zero it follows that

$$\hat{x}_q' = y_q'/\hat{\beta}$$

is an unbiased estimator of x_q'. To determine confidence limits for x_q' we require a special case of a result due to Fieller (1940, 1944) concerning the distribution of the ratio of two normal variates. As y_q' and $\hat{\beta}$ are both normally distributed the result may be applied here.

Let θ represent the variate $y_q'/\hat{\beta}$, then $y_q' - \hat{\beta}\theta$ has zero mean and estimated variance

$$s^2 + \frac{s^2}{n} + \frac{\theta^2 s^2}{s_{x'x'}}$$

where s^2 is an estimate of σ^2 with f degrees of freedom. In this expression $s^2 + s^2/n = (n+1)s^2/n$ is the estimated variance of the single observation $y_q' = y_q - \bar{y}$; $s_{x'x'} = \sum x_i^2$.

It follows from general statistical theory that

$$t^2 = \frac{(y_q' - \hat{\beta}\theta)^2}{\{(n+1)/n\}s^2 + \theta^2 s^2/s_{x'x'}} \tag{7.1}$$

has an F distribution with 1 and f degrees of freedom. In view of the relation between the distribution $F_{1,f}$ and t_f, if we set t^2 equal to the value of $F_{1,f}$ that will just indicate significance at some chosen level α, the $1 - \alpha$ confidence limits for x_q' will be given by the roots of (7.1) regarded as a quadratic in θ, which may be written

$$\theta^2(\hat{\beta}^2 - t^2 s^2/s_{x'x'}) - 2\theta y_q'\hat{\beta} + y_q'^2 - (n+1)t^2 s^2/n = 0,$$

and the roots of this equation, $\tilde{\theta}$, say, are

$$\tilde{\theta} = \frac{y_q'\hat{\beta} \pm ts\sqrt{\{(n+1)\hat{\beta}^2/n + y_q'^2/s_{x'x'} - (n+1)t^2 s^2/(n s_{x'x'})\}}}{\hat{\beta}^2 - t^2 s^2/s_{x'x'}} \tag{7.2}$$

Remembering that $y_q' = \hat{\beta}\hat{x}_q'$ it follows that confidence limits for x_q' may be written, using (7.2) as

$$\tilde{\theta} = \frac{x_q'\hat{\beta}^2 \pm ts\sqrt{\{(n+1)\hat{\beta}^2/n + \hat{\beta}^2 x_q'^2/s_{x'x'} - (n+1)t^2 s^2/(n s_{x'x'})\}}}{\hat{\beta}^2 - t^2 s^2/s_{x'x'}} .$$

Finally, by adding and subtracting $\hat{x}_q' t^2 s^2/s_{x'x'}$, in the numerator, dividing numerator and denominator by $\hat{\beta}^2$ and writing $g = t^2 s^2/\hat{\beta}^2 s_{x'x'}$ and remembering that $x_q' = x_q - \bar{x}$, $y_q' = y_q - \bar{y}$ and $s_{x'x'} = \sum x_i'^2 = \sum (x_i - \bar{x})^2 = s_{xx}$, say, the confidence limits for x_q may be written

$$L = \hat{x}_q + \frac{(x_q - \bar{x})g \pm (ts/\beta)\sqrt{\{(x_q - \bar{x})^2/s_{xx} + (n+1)(1-g)/n\}}}{1-g} . \quad (7.3)$$

If g is small (in practice Finney (1952) has suggested that $g < 0.05$ may be regarded as small), it is reasonably safe to ignore it and work with approximate limits given by putting $g = 0$ in (7.3). In general g will tend to be large if $|\hat{\beta}|$ is small or badly determined, in which case s^2 will tend to be large or s_{xx} will tend to be small, or both. Some complications in interpretation can occur. For example, the roots of (7.2) may turn out to be complex. In this case we must conclude that for the observed y_q, any value of x would be acceptable. This in turn implies that $\hat{\beta}$ is not significantly different from zero. A full discussion of complications is given in Williams (1959, chapter 6) who refers to the limits we have obtained above by their more usual name of *tolerance limits*. Both he and Cox (1968) point out that if it is possible to determine the regression of X on Y (i.e. if we have random variate values of both from which to determine the regression of either variate upon the other) it is a more efficient procedure to estimate a value of X for given Y, by considering the regression of X on Y. We stress that this latter procedure is only valid if the values of X are a random sample of values of that variate.

7.3. Comparison of regression equations

In section 7.1 we mentioned the important concept of reproducibility of results when regression equations are used for description or prediction. We often wish to know if several estimated regression equations can all be regarded as estimates of the same fundamental equation.

We may be interested in studying the effect of different treatments upon a regression line. For example, when observations are made on differently treated units we may wish to know whether the regression lines or planes corresponding to different treatments are identical, or parallel, or whether the coefficients bear no relation to one another.

In other circumstances we may be more interested in questions of concurrence of a number of lines. Again, there are circumstances in which applications of a treatment may cause a sudden change in the parameters of a regression relationship. To deal fully with questions of this type would require a book in itself; however, we illustrate the fundamental ideas of studying such relationships by demonstrating the procedure for a few particular situations. Tests of various hypotheses can often be made using an analysis of variance when the classical regression assumptions of independent normally distributed departures with common variance σ^2 holds.

7.4. Parallel and identical regression lines and planes

Tests for parallelism and identity are given in a number of texts, but as they are important we outline the main results in this section. We consider first the case of a single regressor. We suppose each set of observations to which a line is fitted represents a different treatment, and consider first the hypothesis that all lines are parallel. Suppose there are t treatments. If the regression line for the ith treatment is written

$$y = \alpha_i + \beta_i x$$

the hypothesis of parallelism is that

$$\beta_1 = \beta_2 = \ldots = \beta_t.$$

Identity of the regression lines would require in addition that

$$\alpha_1 = \alpha_2 = \ldots = \alpha_t.$$

As the test for parallelism includes a condition that is also required for identity, it is customary to test for parallelism first, since it is only worth testing for identity after we have accepted a hypothesis of parallelism, i.e. of the lines having a common slope. Basically a test for parallelism requires a comparison of the fit obtained when each line is allowed a different slope with that obtained when all are constrained to have a common slope.

The procedure is to fit t separate lines by the methods of chapter 2. For each line we obtain an estimate $s_i{}^2$, $i = 1, 2, \ldots, t$ of σ^2 with $n_i - 2$ degrees of freedom where n_i is the number of sets of data for the

ith line. A combined estimate of σ^2 with $N-2t$ degrees of freedom where $N = \sum n_i$ is given by

$$s^2 = \frac{\sum (n_i-2)s_i{}^2}{\sum (n_i-2)} = \frac{\sum (n_i-2)s_i{}^2}{N-2t} \qquad (7.4)$$

This is the residual mean square deviation based upon deviations from individual regressions. It will be used as an *error* term in the subsequent analysis of variance.

To fit a common β to all lines we minimize, by the least squares principle,

$$\sum_i \left\{ \sum_k (y_{ik} - \alpha_i - \beta x_{ik})^2 \right\},$$

$k = 1, 2, \ldots, n_i;\; i = 1, 2, \ldots, t.$

Adopting an obvious generalization of our notation s_{xx}, s_{xy} introduced in section 2.6 the estimator of $\hat{\beta}$ is easily shown to be

$$\hat{\beta} = \frac{\sum_i \left\{ \sum_k (y_{ik} - \bar{y}_i)(x_{ik} - \bar{x}_i) \right\}}{\sum_i \left\{ \sum_k (x_{ik} - \bar{x}_i)^2 \right\}} = \frac{\sum_i s_{x_i y_i}}{\sum_i s_{x_i x_i}}.$$

The regression sum of squares accounted for by fitting $\hat{\beta}$ is easily shown to be $\hat{\beta} \sum_i s_{x_i y_i}$ with one degree of freedom. For the individual regressions the sum of squares with t degrees of freedom is $\sum_i \hat{\beta}_i s_{x_i y_i}$; the difference $\sum_i (\hat{\beta}_i - \hat{\beta}) s_{x_i y_i}$ has $t-1$ degrees of freedom and represents the difference in fit between that with individual estimates of slope and that using a common estimator $\hat{\beta}$. If all the true β_i are equal, under the hypothesis of parallelism

$$\frac{\sum (\hat{\beta}_i - \hat{\beta}) S_{x_i y_i}}{(t-1)s^2}$$

will have an F distribution with $t-1$ and $N-2t$ degrees of freedom.

It is often clear from the data that there is no question of all lines being identical, and the test for parallelism is all that is required.

If the hypothesis of parallelism is accepted, a necessary but not sufficient condition for identity is that the means for each line depart

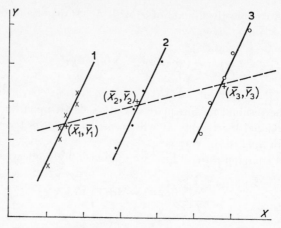

FIG. 7.1. Means collinear but lines 1, 2, 3 not identical

only from a straight line by amounts accounted for in terms of the random departures associated with individuals. Figure 7.1 shows a situation where the condition is clearly not sufficient. If a line can be satisfactorily fitted through the means and has estimated slope $\bar{\beta}$, the hypothesis of identity is only acceptable if $\hat{\beta}$, $\bar{\beta}$ do not differ significantly. The least squares estimator of $\bar{\beta}$ is

$$\bar{\beta} = \frac{\sum\limits_{i=1}^{t} n_i(\bar{x}_i - \bar{x})(\bar{y}_i - \bar{y})}{\sum\limits_{i=1}^{t} n_i(\bar{x}_i - \bar{x})^2} \tag{7.6}$$

where $\bar{x} = \sum n_i\bar{x}_i/N$, $\bar{y} = \sum n_i\bar{y}_i/N$.

The sum of squares for deviations of individual means from the regression line through $(\bar{x}, \bar{y})$ with slope $\bar{\beta}$ has $t-2$ degrees of freedom and is given by

$$\sum_{i=1}^{t} n_i(\bar{y}_i - \bar{y})^2 - \sum_{i=1}^{t} n_i\bar{\beta}(\bar{y}_i - \bar{y})(\bar{x}_i - \bar{x}).$$

The ratio of the corresponding mean square to s^2 will have an F

distribution under the hypothesis that the means of all populations lie on a straight line. If this hypothesis is accepted, all that remains is to compare $\bar{\beta}$ and $\hat{\beta}$.

If they are not significantly different

$$E(\bar{\beta}-\hat{\beta}) = 0$$

and $\mathrm{Var}\ (\bar{\beta}-\hat{\beta}) = s^2 \left\{ \dfrac{1}{\sum n_i(\bar{x}_i - \bar{x})^2} + \dfrac{1}{\sum\limits_i s_{x_i x_i}} \right\} = s_d^2$, say.

Under the null hypothesis

$$t = \frac{|\bar{\beta}-\hat{\beta}|}{s_d}$$

will have a t distribution with $N-2t$ degrees of freedom.

We may also consider tests for parallelism and/or identity of hyperplanes in multiple regression. We here sketch an approach given by Williams (1959, chapter 8). Suppose there are q regressors. Individual regression coefficients fitted to each of t sets of data therefore account for tq degrees of freedom. If common regression coefficients (apart from the constant term) are fitted to all sets of data they will account for q degrees of freedom. The difference with $tq - q = q(t-1)$ degrees of freedom represents departures from parallelism. The ratio of the corresponding mean square to the mean square for departures from individual lines with $N - t(q+1)$ degrees of freedom has an F distribution under the hypothesis of parallelism. This is an obvious generalization of the test for parallelism of lines.

Tests for identity run into difficulty if we try a direct generalization of the methods given for a single regressor. The test as to whether the means lie in a hyperplane generalizes quite easily, but the t test for identity of $\hat{\beta}_i$ and $\bar{\beta}_i$ cannot easily be extended to q coefficients when their estimates are correlated. Williams in fact provides a blanket test using an approach allied to covariance analysis. Details are given in chapter 8 of his book. The procedure may also be looked upon as a special case of more general procedures for testing equality of a group of parameters, or for testing whether certain parameters have specified values. Such problems are discussed by Williams and numerous other writers.

7.5. Concurrent lines

It is sometimes clear that fitted lines are not consistent with identical or parallel population lines, and yet they may reasonably be supposed to be estimates of concurrent lines. Similarly a number of planes may all intersect along one line and hyperplanes in q dimensions may all intersect in the same hyperplane in a lower dimension. In this section we confine our attention to classical least squares regression with a single regressor. In a sense parallel lines may be looked upon as lines concurrent at infinity. Williams has dealt fully with concurrence in section 8.4 of his book, and we only illustrate methods for certain simple cases.

Concurrence is of interest, for example, in an experiment where, for a given value of x, all treatments give, apart from random variation, the same response y, but as x moves away from this fixed value the rate of change of y with respect to x is different for each treatment.

We may, as well as testing a hypothesis of concurrence, wish to estimate the point of concurrence. We consider the case where the x co-ordinate, x_c, say, of the point of concurrence is known. Williams indicates that little modification is needed if x_c has to be estimated from the data, providing this is done in an appropriate manner (see exercise 7.2). When x_c is known, y_c, the ordinate of the point of concurrence, may be estimated from the mean regression line for all the data. We consider here the case where the observed x are the same for all sets of data.

The mean regression line for all the data may then be written

$$y = \bar{y} + \hat{\beta}(x - \bar{x}),$$

and an estimate $\hat{y}_c$ is obtained from this equation as

$$\hat{y}_c = \bar{y} + \hat{\beta}(x_c - \bar{x}).$$

Tests for concurrence can clearly be based on comparison of fits of unconstrained lines with fits constrained to pass through x_c, $\hat{y}_c$. The regression coefficients in the constrained lines are estimated as

$$\beta'_i = \frac{\sum\limits_k (y_{ik} - \hat{y}_c)(x_k - x_c)}{\sum\limits_k (x_k - x_c)^2} \tag{7.7}$$

in an obvious notation; note that we only require one suffix on the x as we assume these to be the same for all sets of data. The sum of squares for concurrent regressions has $t + 1$ degrees of freedom as we have estimated t slopes and also $\hat{y}_c$. If there are n points available to determine each line, the corresponding sum of squares is easily shown to be

$$\frac{\sum_i \left\{ \sum_k y_{ik}(x_k - x_c) \right\}^2}{\sum (x_k - x_c)^2} + \frac{\hat{y}_c^2 nt \sum (x_k - \bar{x})^2}{\sum (x_k - x_c)^2} \tag{7.8}$$

where the first term with t degrees of freedom is associated with the regression through $(x_c, 0)$ and the second term is associated with departures of $\hat{y}_c$ from zero. Williams points out that in most applications it would be more appropriate to partition into a part accounting for the regression through (x_c, y_c) and a part accounting for departures of $\hat{y}_c$ from y_c. Equation (7.8) is the special case with $y_c = 0$. To test significance of departure from a hypothetical y_c, the appropriate sum of squares is clearly

$$\frac{(\hat{y}_c - y_c)^2 nt \sum (x_k - \bar{x})^2}{\sum (x_k - x_c)^2}. \tag{7.9}$$

However, before testing such a hypothesis it is desirable to test whether there are significant departures from concurrence, given x_c. A sum of squares for departures is given by subtracting (7.8) from the sum of squares attributable to individual treatment means $\bar{y}_1, \bar{y}_2, \ldots, \bar{y}_t$ and regression coefficients $\hat{\beta}_1, \hat{\beta}_2, \ldots, \hat{\beta}_t$ with $2t$ degrees of freedom in all. The difference with $t - 1$ degrees of freedom may be tested against residual variation about individual regression lines. This has $t(n - 2)$ degrees of freedom and the F test is applied in the usual way. If the result is significant it is not worth proceeding further. If it is not significant, hypotheses about y_c may be tested using appropriate sums of squares of the type (7.9). Williams, for instance, suggests that the first term in (7.8) may be partitioned into a sum of squares with $t - 1$ degrees of freedom representing differences between concurrent regression lines, and one degree of freedom representing mean regression through the point of concurrence.

He suggests practical ways of calculating the various sums of squares in an analysis of variance, and for partitioning these. Methods most suitable for desk machines may not always be appropriate for an electronic computer.

7.6. Two phase regressions

A problem closely related to concurrence of regression lines occurs when a regression line undergoes an abrupt change of slope as x passes through a critical value γ, say, i.e. for $x < \gamma$,

$$y = \alpha_1 + \beta_1 x, \tag{7.10}$$

while for $x > \gamma$,

$$y = \alpha_2 + \beta_2 x. \tag{7.11}$$

When we are dealing with more than one set of data, each set corresponding to a different treatment, we may want to know whether it is reasonable to hypothesize the same γ for each treatment. As in the case of concurrence, various hypotheses can be envisaged and tested. We illustrate the principles involved for the case of fitting lines (7.10) and (7.11) where these lines are constrained to intersect on the line $x = \gamma$. A number of related hypotheses and tests are discussed by Sprent (1961).

One situation where sudden changes in slope occur is that in which x is an increasing function of time (a class of functions that includes time itself) and at time t_c a treatment is applied. This may well affect the slope of the regression line—perhaps immediately, or sometimes only after a time lag.

It is convenient to call the line applicable to values of x less than γ the first phase of the regression and that for x greater than γ the second phase. For the case of one such two-phase regression with the constraint that the lines intersect on the line $x = \gamma$ we note that at the intersection

$$y = \alpha_1 + \beta_1 \gamma = \alpha_2 + \beta_2 \gamma.$$

Hence the coefficients must satisfy the constraint

$$\alpha_2 - \alpha_1 + \gamma(\beta_2 - \beta_1) = 0. \tag{7.12}$$

Introducing an undetermined multiplier λ, the function to be minimized to obtain a least-squares solution subject to this constraint is

$$\sum\{\sum(y_{rk}-\alpha_r-\beta_r x_{rk})^2\}+2\lambda\{\alpha_2-\alpha_1+\gamma(\beta_2-\beta_1)\}$$

where $r = 1, 2$, corresponding to the two phases and k runs from 1 to n_r, n_r being the number of observed points in the rth phase. The resulting normal equations may be written as (7.12) together with

$$\alpha_r = \bar{y}_r-\beta_r\bar{x}_r+(-)^{r-1}\lambda/n_r, \tag{7.13}$$

$$\sum_k(y_{rk}x_{rk}-\alpha_r x_{rk}-\beta_r x_{rk}{}^2)+(-)^{r-1}\lambda\gamma = 0. \tag{7.14}$$

Subtracting the equations corresponding to the two values of r from one another in (7.13) and using (7.12) gives

$$\lambda = w\{\bar{y}_2-\beta_2(\bar{x}_2-\gamma)-[\bar{y}_1-\beta_1(\bar{x}_1-\gamma)]\} \tag{7.15}$$

where $w = n_1 n_2/(n_1+n_2)$. Substitution for α_1, α_2 and λ as given by (7.13) and (7.15) in (7.14) gives as normal equations for the estimation of β_1, β_2 in an obvious notation

$$\begin{aligned}[s_{x_1x_1}+w(\bar{x}_1-\gamma)^2]\beta_1-w(\bar{x}_1-\gamma)(\bar{x}_2-\gamma)\beta_2 \\ = s_{x_1y_1}-w(\bar{y}_2-\bar{y}_1)(\bar{x}_1-\gamma)\end{aligned} \tag{7.16}$$

$$\begin{aligned}-w(\bar{x}_1-\gamma)(\bar{x}_2-\gamma)\beta_1+[s_{x_2x_2}+w(\bar{x}_2-\gamma)^2]\beta_2 \\ = s_{x_2y_2}+w(\bar{y}_2-\bar{y}_1)(\bar{x}_2-\gamma).\end{aligned} \tag{7.17}$$

These equations may be solved to give estimators $\hat{\beta}_1$, $\hat{\beta}_2$ of β_1, β_2; and $\hat{\alpha}_1$, $\hat{\alpha}_2$, the estimators of α_1, α_2, and λ are then obtainable from (7.13) and (7.15). We may compare the fit constrained to pass through a point on the line $x = \gamma$ with the fit of an unconstrained pair of lines. For the unconstrained pair of lines there are four independent parameters to be estimated. For the constrained line there are only three independent parameters in view of the constraint, (7.12). It is not difficult to show that the one degree of freedom representing departures from concurrence has a sum of squares that is obtained by subtracting

$$2\tilde{\beta}_1 s_{x_1y_1} - \tilde{\beta}_1{}^2 s_{x_1x_1} + 2\tilde{\beta}_2 s_{x_2y_2} - \tilde{\beta}_2{}^2 s_{x_2x_2} - \lambda^2/w \qquad (7.18)$$

from the unconstrained regression sum of squares, $\tilde{\beta}_1 s_{x_1y_1} + \tilde{\beta}_2 s_{x_2y_2}$, where $\tilde{\beta}_1$, $\tilde{\beta}_2$ are the regression coefficients for the unconstrained model.

This sum of squares is analogous to the sum of squares for departures from concurrency discussed in section 7.5, i.e. it is appropriate to a test for departures from concurrency with the given abscissa.

7.7. Residuals

The electronic computer has provided an excellent tool for computing the departures of each point from a fitted line. These are called the *residuals*. In the pre-computer era it was customary to look only at the residual sum of squares, usually obtained by subtraction of the regression sum of squares from the total. Occasionally components of this sum of squares were examined, e.g. if several lines were being fitted to different sets of data, it was sometimes of interest to see how much each contributed to the residual sum of squares, and more particularly, by examining the mean square corresponding to each such component any heterogeneity of variance could be detected, i.e. evidence that σ^2 was not the same for each set of data.

For example, Sprent (1967) fitted linear regressions of a dental measurement on age for each of 16 boys using data given by Potthoff and Roy (1964). The residual sums of squares, each with two degrees of freedom, had the values

4.7,	2.2,	4.0,	1.6,	4.0,	0.1,	1.5,	5.9
42.2,	1.2,	0.6,	2.2,	7.2,	2.2,	1.9,	1.5

The value 42.2 printed in italics contributes more than 50 per cent of the total residual sum of squares, so some doubt is immediately cast upon the assumption that the departures all have the same distribution for each set of data, and any tests made with a pooled estimate of σ would be suspect. In section 2.7 we gave a test, known as Bartlett's test, for deciding whether a number of independent estimates of variance are in fact estimating a common variance. With the numerical data just discussed, the large residual 42.2 at least suggests looking carefully at the observations. The measurements for this subject certainly looked most peculiar for a growing boy, and their peculiarity was attributed by Potthoff and Roy to the fact that what was recorded was distance between two growing points not fixed relative to one another.

We can get a lot of information about adequacy of our models by studying the residuals associated with each point. Methods for studying these are given by Draper and Smith (1966, Chapter 3). A more general study of residuals, not limited to linear regression models, has been made by Cox and Snell (1968).

For the linear regression model the residual is $e_i = y_i - E'(y_i)$ where y_i is the observed value and $E'(y_i)$ the estimator of $E(y)$ is,

$$E'(y_i) = \hat{\beta}_0 + \hat{\beta}_1 x_{1i} + \hat{\beta}_2 x_{2i} + \ldots + \hat{\beta}_q x_{qi}. \qquad (7.19)$$

Given n observed points, we may consider a vector of residuals, **e**, such that

$$\mathbf{y} = \mathbf{X}\hat{\boldsymbol{\beta}} + \mathbf{e}$$

where **y**, **X** and $\hat{\boldsymbol{\beta}}$ are as defined in section 5.2. If the number of estimated parameters is small compared to the number of observations it is reasonable to expect that **e** will have similar properties to **ε** in the model

$$\mathbf{Y} = \mathbf{X}\boldsymbol{\beta} + \boldsymbol{\varepsilon}.$$

That this need not be so if q in (7.19) is of the same order as n, the number of points observed, is evident by considering the extreme case where we determine n parameters from n points. Then the residuals are all zero whatever the distribution of **ε** as an exact fit is obtained. This does not, of course, imply that any further observations will satisfy the estimated regression equation exactly.

Even if the elements of **ε** are independent $N(0, \sigma^2)$, the elements of **e** will be correlated with one another and have a joint singular normal distribution and satisfy the constraint $\sum e_i = 0$. It is of some

109

interest to note that the correlation between residuals depends entirely upon the matrix $\mathbf{X}$ of observations, and not upon σ^2. To show this, we note that

$$\begin{aligned} \mathbf{e} = \mathbf{y} - E'(\mathbf{y}) &= \mathbf{y} - \mathbf{X}\hat{\boldsymbol{\beta}} \\ &= \mathbf{y} - \mathbf{X}(\mathbf{X}'\mathbf{X})^{-1}\mathbf{X}'\mathbf{y}. \end{aligned}$$

Since $E(\mathbf{y}) = \mathbf{X}\boldsymbol{\beta}$, it follows that

$$\begin{aligned} \mathbf{e} - E(\mathbf{e}) &= \mathbf{y} - \mathbf{X}(\mathbf{X}'\mathbf{X})^{-1}\mathbf{X}'\mathbf{y} - \{\mathbf{X}\boldsymbol{\beta} - \mathbf{X}(\mathbf{X}'\mathbf{X})^{-1}\mathbf{X}'\mathbf{X}\boldsymbol{\beta}\} \\ &= \{\mathbf{I} - \mathbf{X}(\mathbf{X}'\mathbf{X})^{-1}\mathbf{X}'\}(\mathbf{y} - \mathbf{X}\boldsymbol{\beta}) \\ &= \{\mathbf{I} - \mathbf{X}(\mathbf{X}'\mathbf{X})^{-1}\mathbf{X}'\}\boldsymbol{\varepsilon}, \end{aligned}$$

whence, writing Var $(\mathbf{e})$ for the covariance matrix of $\mathbf{e}$

$$\mathrm{Var}(\mathbf{e}) = \{\mathbf{I} - \mathbf{X}(\mathbf{X}'\mathbf{X})^{-1}\mathbf{X}'\} E(\boldsymbol{\varepsilon}\boldsymbol{\varepsilon}') \{\mathbf{I} - \mathbf{X}(\mathbf{X}'\mathbf{X})^{-1}\mathbf{X}'\}'$$

where, for the classical least-squares regression assumptions, $E(\boldsymbol{\varepsilon}\boldsymbol{\varepsilon}') = \sigma^2 \mathbf{I}$. Thus

$$\begin{aligned} \mathrm{Var}(\mathbf{e}) &= \{\mathbf{I} - \mathbf{X}(\mathbf{X}'\mathbf{X})^{-1}\mathbf{X}'\} \{\mathbf{I} - \mathbf{X}(\mathbf{X}'\mathbf{X})^{-1}\mathbf{X}'\} \sigma^2 \\ &= \{\mathbf{I} - \mathbf{X}(\mathbf{X}'\mathbf{X})^{-1}\mathbf{X}'\} \sigma^2. \end{aligned} \tag{7.20}$$

Note that $\mathbf{I} - \mathbf{X}(\mathbf{X}'\mathbf{X})^{-1}\mathbf{X}' = \mathbf{R}$ has the property of *idempotency*, i.e. $\mathbf{R} = \mathbf{R}^2$. The correlation between two elements e_i, e_j of $\mathbf{e}$ is given by

$$\rho_{ij} = \frac{\mathrm{Cov}(e_i, e_j)}{\{\mathrm{Var}(e_i)\mathrm{Var}(e_j)\}^{1/2}}. \tag{7.21}$$

From the form of (7.20) it is clear that σ^2 cancels between numerator and denominator in (7.21), and thus the correlation depends only upon $\mathbf{X}$.

Cox and Snell (1968) list the following ways, (1 to 6) in which departures from classical least squares regression models may occur, and where a study of residuals may be of use. We have already met some of these situations in earlier chapters.

1. *There may be outliers.* Some caution is needed in testing for these. Generally speaking it is only safe to reject an observation as an outlier if there is strong non-statistical evidence that it is abnormal. Enquiries as to how the data were obtained may indicate, for

instance, that the suspect observation was obtained on the day that a fault in the recording instrument was reported. If the data have been obtained by manipulation of the original observations (e.g. by transformation to logarithms, computation of a volume from linear measurements, etc.) an error may be found in the manipulation. With small numbers of observations it is virtually impossible to tell whether or not a single observation is an outlier. Such a situation is illustrated for six observations in figure 7.2. If the observation

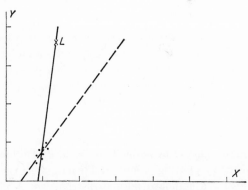

FIG. 7.2. A doubtful outlier

labelled L in that figure is omitted, the dotted regression line is obtained by a least-squares fit to the remaining data. If L is included, the solid line is obtained. Each provides a reasonable fit to its own data, but if the observation L had not been taken, very different conclusions would have been made about the regression of Y on x compared to those that would have been arrived at with L included. There may be non-statistical grounds for excluding L, but if not, an experimenter faced with a situation like that in figure 7.2 would be well advised to try and obtain additional observations —and if possible to extend the range of x values included.

2. *There may be a further useful regressor, x_{q+1}, say, omitted from* (7.19). This may sometimes be detected by plotting residuals against levels of this factor. A difficulty here is that the values of x_{q+1} may not be known. If they had been, it is probable that x_{q+1} would have been tried as a regressor.

111

3. *There may be non-linear regression on a regressor already in* (7.16). This may be detected by plotting the residuals against the regressor values, when a curvilinear relationship may be detected.

4. *There may be correlations between departures.* These are common between observations on the same unit adjacent in time, and can sometimes be detected by a plot of residuals ordered in time, or from a scatter diagram of pairs of residuals suspected of being correlated.

5. *There may be heterogeneity of variance.* This can often be detected by plotting residuals, or squares of residuals, against factors thought to affect variance. For example, with a single regressor a plot of residuals against regressor values, or their squares, may be informative. A plot against $E'(y_i)$, the fitted values, sometimes proves illuminating.

6. *The departures may not be normally distributed.* This may sometimes be detected by plotting ordered residuals against the expected order statistics for a standard normal distribution. A plot of the frequency curve of the observed values of e_i will also indicate if their distribution looks like a random sample from a normal distribution with zero mean.

A form of plot sometimes useful for picking up outliers or departures from normality is the *normal plot* or *half normal plot* on probability paper proposed by Daniel (1959). If the classical regression assumptions hold, the plotted points should lie approximately on a straight line.

It is possible to associate tests of hypotheses of a formal nature with the above graphical techniques.

There is little evidence that the correlation between residuals seriously upsets the graphical techniques described above.

7.8. Regression analysis of designed experiments

In the regression theory considered so far we have assumed implicitly, even if not explicitly, that the number of observed points was greater than or equal to $q+1$, where q was the number of parameters to be fitted. When n is less than q, clearly there is no unique regression equation. The most extreme example of this occurs for a single regressor, where it is clearly impossible to fit a unique line $y = \alpha + \beta x$ through only one point! In more general cases attempts to fit more parameters than there are observational points results in the matrix $\mathbf{X'X}$ becoming singular, i.e. it has a zero determinant and $(\mathbf{X'X})^{-1}$ does not exist.

The condition $n \geq q+1$ is not a sufficient condition for $(\mathbf{X'X})^{-1}$ to exist. A counter example has already been given in section 5.2. The example there is, statistically speaking, a pathological one, in that perfect correlation is unlikely to occur between observational regressors.

There is, however, one important field of application of regression theory where singularity of $\mathbf{X'X}$ is the rule rather than the exception. This is in experimental design and analysis. The relationship between the analysis of variance of standard experimental designs and regression analysis was recognized over thirty years ago, and several books on experimental design, e.g. Kempthorne (1952), make regression analysis the starting point for their work on experimental design. As we assume our readers to have a basic knowledge of statistical theory, we shall merely illustrate this relationship in terms of the well-known randomized block design. The example brings out the way that certain characteristics of experimental design reflect themselves in the form of $\mathbf{X'X}$ and $(\mathbf{X'X})^{-1}$.

In a randomized block experiment where each treatment occurs once in each block, the yield y_{ij} of the plot in the ith block receiving treatment j may be written

$$y_{ij} = \beta_i + \tau_j + \varepsilon_{ij}.$$

There are $r+t$ parameters $\beta_1, \beta_2, \ldots, \beta_r, \tau_1, \tau_2, \ldots, \tau_t$, in this model for a randomized block experiment with t treatments in r blocks. It is well known to users of the analysis of variance, whether or not they regard it as a regression model, that not all these parameters can be estimated uniquely. To obtain unique estimates it is necessary to apply constraints. In practice it is in fact usual to introduce a further parameter μ and two constraints $\sum \beta_i = 0$, $\sum \tau_j = 0$, when the model is rewritten

$$y_{ij} = \mu + \beta_i + \tau_j + \varepsilon_{ij}.$$

This may be further rewritten as a multiple regression model in the form

$$y_{ij} = \mu x_0 + \beta_1 x_1 + \beta_2 x_2 + \ldots + \beta_r x_r + \tau_1 z_1 + \ldots + \tau_t z_t + \varepsilon_{ij} \quad (7.22)$$

where $x_1, x_2, \ldots, x_r, z_1, z_2, \ldots, z_t$ are a set of $r+t$ regressors such that for a given y_{ij}, $x_i = 1$, $x_k = 0$, $k \neq i$, and $z_j = 1$, $z_l = 0$, $l \neq j$, while $x_0 \equiv 1$. We make the classical regression assumptions about ε_{ij}.

If we proceed by the methods of section 5.2, $\mathbf{X}$ corresponding to yields in a standard order

$$y_{11}, y_{12}, \ldots, y_{1t}, y_{21}, \ldots, y_{2t}, \ldots, y_{r1}, \ldots, y_{rt}$$

may be written down. For example, if $r = 3$ and $t = 4$, $\mathbf{X}$ is

$$\begin{pmatrix}
1 & 1 & 0 & 0 & 1 & 0 & 0 & 0 \\
1 & 1 & 0 & 0 & 0 & 1 & 0 & 0 \\
1 & 1 & 0 & 0 & 0 & 0 & 1 & 0 \\
1 & 1 & 0 & 0 & 0 & 0 & 0 & 1 \\
1 & 0 & 1 & 0 & 1 & 0 & 0 & 0 \\
1 & 0 & 1 & 0 & 0 & 1 & 0 & 0 \\
1 & 0 & 1 & 0 & 0 & 0 & 1 & 0 \\
1 & 0 & 1 & 0 & 0 & 0 & 0 & 1 \\
1 & 0 & 0 & 1 & 1 & 0 & 0 & 0 \\
1 & 0 & 0 & 1 & 0 & 1 & 0 & 0 \\
1 & 0 & 0 & 1 & 0 & 0 & 1 & 0 \\
1 & 0 & 0 & 1 & 0 & 0 & 0 & 1
\end{pmatrix}$$

Clearly $\mathbf{X}$ is a highly patterned matrix. $\mathbf{X'X}$ is easily found to be

$$\begin{pmatrix}
12 & 4 & 4 & 4 & 3 & 3 & 3 & 3 \\
4 & 4 & 0 & 0 & 1 & 1 & 1 & 1 \\
4 & 0 & 4 & 0 & 1 & 1 & 1 & 1 \\
4 & 0 & 0 & 4 & 1 & 1 & 1 & 1 \\
3 & 1 & 1 & 1 & 3 & 0 & 0 & 0 \\
3 & 1 & 1 & 1 & 0 & 3 & 0 & 0 \\
3 & 1 & 1 & 1 & 0 & 0 & 3 & 0 \\
3 & 1 & 1 & 1 & 0 & 0 & 0 & 3
\end{pmatrix}$$

The determinant of the above matrix is easily seen to be zero, as the sum of the second, third and fourth rows equals the first row. Thus we cannot calculate $(\mathbf{X}'\mathbf{X})^{-1}$. If we now apply the constraints $\sum\beta_i = \sum\tau_j = 0$ we can overcome this difficulty by seeking a minimum subject to these constraints. A naive approach would be to use the equations of constraint to eliminate two parameters from (7.22), e.g. β_r and τ_t, whence (7.22) can be written

$$y_{ij} = \mu_0 x_0' + \beta_1 x_1' + \ldots + \beta_{r-1} x_{r-1}' + \tau_1 z_1' + \ldots + \tau_{t-1} z_{t-1}' + \varepsilon_{ij},$$

where $x_0' = x_0 \equiv 1$, $x_i' = x_i - x_r$, $z_j' = z_j - z_t$, $i = 1, 2, \ldots, r-1$; $j = 1, 2, \ldots, t-1$. In terms of the primed regressors $\mathbf{X}$ is

$$\begin{pmatrix}
1 & 1 & 0 & 1 & 0 & 0 \\
1 & 1 & 0 & 0 & 1 & 0 \\
1 & 1 & 0 & 0 & 0 & 1 \\
1 & 1 & 0 & -1 & -1 & -1 \\
1 & 0 & 1 & 1 & 0 & 0 \\
1 & 0 & 1 & 0 & 1 & 0 \\
1 & 0 & 1 & 0 & 0 & 1 \\
1 & 0 & 1 & -1 & -1 & -1 \\
1 & -1 & -1 & 1 & 0 & 0 \\
1 & -1 & -1 & 0 & 1 & 0 \\
1 & -1 & -1 & 0 & 0 & 1 \\
1 & -1 & -1 & -1 & -1 & -1
\end{pmatrix}$$

and $\mathbf{X}'\mathbf{X}$ becomes

$$\begin{pmatrix}
12 & 0 & 0 & 0 & 0 & 0 \\
0 & 8 & 4 & 0 & 0 & 0 \\
0 & 4 & 8 & 0 & 0 & 0 \\
0 & 0 & 0 & 6 & 3 & 3 \\
0 & 0 & 0 & 3 & 6 & 3 \\
0 & 0 & 0 & 3 & 3 & 6
\end{pmatrix}$$

The determinant of this matrix is not zero and it is easily shown that the inverse is

$$\frac{1}{12}\begin{pmatrix}
1 & 0 & 0 & 0 & 0 & 0 \\
0 & 2 & -1 & 0 & 0 & 0 \\
0 & -1 & 2 & 0 & 0 & 0 \\
0 & 0 & 0 & 3 & -1 & -1 \\
0 & 0 & 0 & -1 & 3 & -1 \\
0 & 0 & 0 & -1 & -1 & 3
\end{pmatrix}$$

The estimators of μ, β_1, β_2, τ_1, τ_2, τ_3 are given by $(\mathbf{X'X})^{-1}\mathbf{X'y}$, i.e.

$$
\begin{pmatrix} \hat{\mu} \\ \hat{\beta}_1 \\ \hat{\beta}_2 \\ \hat{\tau}_1 \\ \hat{\tau}_2 \\ \hat{\tau}_3 \end{pmatrix} = \frac{1}{12} \begin{pmatrix} 1 & 0 & 0 & 0 & 0 & 0 \\ 0 & 2 & -1 & 0 & 0 & 0 \\ 0 & -1 & 2 & 0 & 0 & 0 \\ 0 & 0 & 0 & 3 & -1 & -1 \\ 0 & 0 & 0 & -1 & 3 & -1 \\ 0 & 0 & 0 & -1 & -1 & 3 \end{pmatrix} \begin{pmatrix} Y_{..} \\ Y_{1.} - Y_{3.} \\ Y_{2.} - Y_{3.} \\ Y_{.1} - Y_{.4} \\ Y_{.2} - Y_{.4} \\ Y_{.3} - Y_{.4} \end{pmatrix} \quad (7.23)
$$

where $Y_{..} = \sum_i \sum_j y_{ij}$, $Y_{i.} = \sum_j y_{ij}$, $Y_{.j} = \sum_i y_{ij}$.

It is clear from the position of the zero elements in $(\mathbf{X'X})^{-1}$ that μ, (β_1, β_2) and (τ_1, τ_2, τ_3) form systems of parameters that are orthogonal to one another, e.g. any β is uncorrelated with any τ, and they may be estimated separately and the covariance of the estimates is zero. Three degrees of freedom for treatments correspond to the fitting of $\hat{\tau}_1$, $\hat{\tau}_2$, $\hat{\tau}_3$ and the two degrees of freedom for blocks correspond to the fitting of $\hat{\beta}_1$, $\hat{\beta}_2$.

From (7.23) we easily get

$$\hat{\mu} = Y_{..}/12,$$

$$\hat{\beta}_1 = (2Y_{1.} - Y_{2.} - Y_{3.})/12 = Y_{1.}/4 - Y_{..}/12,$$

since $Y_{1.} + Y_{2.} + Y_{3.} = Y_{..}$.

Similarly

$$\hat{\beta}_2 = Y_{2.}/4 - Y_{..}/12,$$

and since $\beta_3 = -\beta_1 - \beta_2$ and their estimators also satisfy this constraint,

$$\hat{\beta}_3 = Y_{3.}/4 - Y_{..}/12.$$

Also

$$\hat{\tau}_j = Y_{.j}/3 - Y_{..}/12, j = 1, 2, 3, 4.$$

The above approach treats estimation of parameters in a randomized block design as a regression problem. To those familiar only with simple designs such as randomized blocks, writing down the above $\mathbf{X}$ matrix and the deductions from it may seem an over-elaborate way of minimizing

$$\sum_{i,j} (y_{ij} - \mu - \beta_i - \tau_j)^2$$

subject to constraints $\sum \beta_i = \sum \tau_j = 0$.

However, the regression approach brings out clearly the well known orthogonality property of blocks and treatments in a randomized block design and also clarifies the degrees of freedom associated with each. The above exposition of the regression approach was naive because it destroys the symmetry of the problem. This could have been retained by introducing Lagrange multipliers and working with an *augmented* matrix to estimate all β_i, τ_j and the multipliers. Such an approach has advantages in so far as it leads more directly to the covariance matrix for all parameters.

We shall not investigate a number of important problems such as the determination of what functions of the parameters are estimable, what form the constraints may take, and how design of an experiment influences the form of $(X'X)^{-1}$. A large part of the study of modern experimental designs is concerned with these questions. One of the aims of good experimental design is to obtain a pattern of $(X'X)^{-1}$ or some sub-matrix thereof, or a matrix derivable therefrom, of a form that ensures that the design will concentrate the information on certain parameters or functions of them that are of interest, i.e. so that the estimates of these will have small standard errors.

Some of the difficulties inherent in the singularity of $(X'X)$ without constraints, and the problem of deciding what constraints are needed and what functions are estimable, have been alleviated by the concept of *pseudo inverses*. We do not pursue this subject further here, but the theory of analysis of variance using pseudo inverses is given by Searle (1966, chapter 10). Beside this text, and that by Kempthorne already cited, further details of the theoretical aspects of experimental design are discussed by Tocher (1952), Pearce (1963), Rao (1965b) and Nelder (1965). The more practical aspects are also discussed by Pearce (1965) and Cochran and Cox (1957). Theoretical and practical aspects of multiple regression with a singular matrix have been discussed recently by Healy (1968).

7.9. Analysis of covariance

The reader will have noted that the regression model in section 7.8 involved regressors that are determined by the experimental design, and are neither random observations nor selected values of observable variables chosen for convenience. There are some experimental

117

situations where we can conveniently introduce a model that includes regressors determined by the design and also regressors that are observed variates or variables. Such a model can be used to advantage in an agricultural experiment, for example, when, as well as measuring yield on any plot, it is possible to make a measurement highly correlated with yield but which is unaffected by the applied treatments. For example, potato yield may be affected by the level of blight infestation on experimental plots. If it can be assumed (and this may be a very big assumption in practice) that level of blight infestation is independent of treatments applied, a measure of the level of blight infestation may be introduced into the regression equation as an additional regressor (usually called a covariate by practitioners of the technique). We do not give details, but merely point out that the additional variate is of use if it reduces the residual mean square significantly. The covariate will not in general be orthogonal to blocks and treatments, so that block and treatment parameters with one or more covariates present will not be the same as those with no covariates present. In theory this means that all parameters and sums of squares must be recomputed if a covariate is introduced. A number of practical techniques exist to "adjust" parameters and sums of squares without undue additional computation. The procedure is usually carried out under the title of *analysis of covariance*.

When more than one covariate is used we speak of multiple covariance. Theoretically there is no reason why covariates that are affected by treatments should not be used, but interpretation of the parameters then becomes difficult. For a detailed discussion of covariance the reader could do little better than consult the special covariance issue of *Biometrics*, listed in the references under Cochran (1957).

7.10. Linear transformations of the dependent variate

Finally in this chapter we mention a technique that has proved useful in growth studies. We have already pointed out that given a set of n values (x_i, y_i) it is always possible to find a polynomial of degree not greater than $n-1$ that passes through all these points. We may therefore replace the original set of n values of Y by the coefficients $\beta_0, \beta_1, \ldots, \beta_{n-1}$ without loss of information. Alternatively, we may replace them by a set of orthogonal polynomial coefficients $\gamma_0, \gamma_1, \ldots, \gamma_{n-1}$, again without loss of information.

If the growth curve of an organism is very nearly quadratic or cubic in time, most of the information about the curve is concentrated in the first few coefficients, and it may suffice to perform an analysis on these alone. The technique was used effectively by Wishart (1938) in a classic study on the growth of pigs, and in a somewhat more sophisticated form by Leech and Healy (1959). It has also been used by Rao (1965a), Sprent (1967) and Pearson and Sprent (1968) with varying degrees of success for summarizing and comparing trends in data. Certain more general test functions can be considerably simplified when applied to orthogonal polynomial coefficients in the light of orthogonality properties.

Exercises

7.1. In the multiple regression equation with q variates how would one test the hypothesis that a sub-set of the coefficients $\beta_1, \beta_2, \ldots, \beta_r$ were all zero? How would one test the hypothesis that $\beta_i = \beta_{0i}$, $i = 1, 2, \ldots, r$, where β_{0i} are specified values? (Kempthorne, (1952, chapter 5)).

7.2. Examine the problem of testing concurrence when the abscissa, x_c, of the point of concurrence has to be estimated from the data.
(Williams, (1959, chapter 8)).

7.3. In section 7.5 various partitionings of the sum of squares (7.8) are suggested. Investigate appropriate ones for hypotheses likely to be of interest. (Williams, (1959, chapter 8)).

7.4. Suppose that instead of testing the hypothesis that the abscissa of the change-over point for a two-phase regression discussed in section 7.6 is γ, we have a situation where observations in each phase are used to fit regression lines to the two phases independently; then where the lines intersect will give an estimate of γ, i.e.

$$\hat{\gamma} = (\hat{\alpha}_2 - \hat{\alpha}_1)/(\hat{\beta}_2 - \hat{\beta}_1).$$

Show how we may use Fieller's result discussed in section 7.2 to obtain confidence limits for γ. (Sprent, 1961).

E

7.5. Examine the linear model appropriate to a 4×4 Latin square after the manner used for a randomized blocks design in section 7.8.

7.6. Explain why, in the analysis of covariance, even if the covariate is not affected by treatments, it is not in general valid to perform an analysis of variance on an adjusted variate $y - \hat{\beta}x$ where $\hat{\beta}$ is estimated from the error line in an analysis of covariance.

(Cochran, (1957)).

CHAPTER 8

Law-like Relationships in Practice

8.1. Number and type of relationships

In sections 6.1 and 6.4 we pointed out that there may often be more than one linear relationship between p variables, and that such relationships determine a sub-space of lower dimension. We also pointed out that with more than one linear relationship, the individual relationships were not unique.

In this section we consider two numerical examples. The first provides an application of the methods developed in section 6.4 and the related concepts of canonical analysis discussed in section 6.5. The second example shows how a number of seemingly different results using different approaches may be reconciled—at least when departures from true values satisfying the relationships are not too marked.

Example 8.1. The data in table 8.1 represent measurements of crop, wood growth and girth increment on apple trees. These measurements were made at East Malling Research Station during the years 1954–1958 on trees of the variety Cox's Orange Pippin. Similar records were in fact taken on a much larger plantation, but for illustrative purposes six trees (not the same six each year) were selected at random for each season, and the study here is confined to these limited records. In these circumstances a model that assumes departures are independent from year to year seems appropriate. With a further assumption of homogeneity of the within years covariance matrix from year to year, the model is precisely that envisaged in section 6.4. An estimate W_1 of W is obtained from replication within each year.

It is the author's opinion that many approaches to the analysis of such multivariate data are far too casual; techniques that are

TABLE 8.1. Log crop (x_1), log extension wood growth (x_2), log girth increment (x_3) for Cox's Orange Pippin apple trees.

Year	x_1	x_2	x_3
	1·03	3·44	0·49
	1·24	3·62	0·51
1954	1·12	3·42	0·43
	1·05	3·47	0·43
	1·11	3·28	0·48
	0·54	3·42	0·52
Means	1·01	3·44	0·48
	0·94	3·37	0·69
	0·76	3·33	0·68
1955	1·32	3·11	0·62
	0·97	3·24	0·64
	1·32	3·13	0·54
	1·41	2·90	0·49
Means	1·12	3·18	0·61
	1·92	3·97	0·56
	1·97	3·97	0·52
1956	1·96	3·79	0·43
	1·91	4·04	0·52
	1·89	3·94	0·49
	1·97	3·95	0·51
Means	1·94	3·94	0·51
	1·68	3·81	0·32
	1·56	4·07	0·46
1957	1·90	4·18	0·51
	1·91	3·78	0·23
	1·51	3·95	0·40
	1·90	4·10	0·57
Means	1·74	3·98	0·41
	2·19	4·06	0·63
	2·24	4·04	0·57
1958	2·10	4·07	0·59
	2·18	4·11	0·64
	2·18	3·93	0·59
	2·15	4·20	0·64
Means	2·17	4·07	0·61

only vaguely relevant are applied in the hope that something interesting will result. In exploratory investigations where little is known about the theory of a process under investigation such an approach may be illuminating, often suggesting a hypothesis that can be examined in more detailed experiments; but when there is some biological, chemical, physical, engineering or other theory as a background it seems more profitable to formulate hypotheses of interest and then decide upon the statistical technique that will weigh the evidence for or against these hypotheses. The observations recorded in table 8.1 were taken as part of a general study of growth and development of apple trees, and one point of interest was a more precise determination of the nature of the known balance between reproductive growth (crop) and vegetative growth (measured by wood growth and girth increment). Although the study was concerned with the elucidation of growth mechanisms, commercial aspects of balance between profitable crop and vegetative growth are of importance. If the biology of a tree can be altered so that it puts a greater proportion of its growth into crop, this is clearly of interest to a fruit grower. Livestock farming has already reached the factory stage, and concentrated production of horticultural crops is being given attention, and certainly raises fewer ethical problems than intensive rearing of livestock.

Determination of a linear relationship between growth measurements will not in itself tell us how to convert more of our growth into reproductive form, but it may help to elucidate the nature of the balance between growing parts and indicate to a physiologist which aspects of growth he should try to alter to get a desired effect such as more crop per unit weight of tree or per unit length of branch. For example, if an analysis showed a linear relationship between girth increment and crop, more or less independent of wood growth, this would indicate that decreasing girth increment may increase crop. It should be pointed out that there is a distinct danger that any relationship that may be found will not extrapolate to values of the variables outside the range of observations. The term wood growth used above refers to extension growth of shoots, and girth increment refers to increment in the circumference of the trunk of a tree at a fixed marker.

In examining whether there is a balance between logarithms of vegetative and reproductive growth measurements, biometricians working on apple trees are using their prior knowledge based on observational experience that there is a tendency for trees of certain varieties to put on more vegetative growth when they carry a light crop than when they carry a heavy one. Extension growth and trunk thickening are the two main types of vegetative growth. Logarithms of the measurements are used as it is believed that volume or area rather than linear measurements may be important, and also that there may be some kind of joint proportionality of crop upon the two types of vegetative growth. Our measured variates are looked upon as being composed of (mathematical) variables with added random departures that in any year are assumed to be distributed $N(\mathbf{0}, \mathbf{W})$ and to be independent from year to year. We have discussed the background in some detail to indicate that there are definite reasons for using logarithms of the data and for seeking linear relationships using the model of section 6.4.

From the data in table 8.1, $\mathbf{B}$ may be computed as the matrix of sums of squares and products between years and $\mathbf{W}_1$, the estimate of $\mathbf{W}$, as the pooled within years covariance matrix. This estimate of $\mathbf{W}$ has 25 degrees of freedom. An equation corresponding to (6.5) may now be written down with $\mathbf{W}_1$ in place of $\mathbf{W}$. Latent roots and vectors may be obtained by methods given by Seal (1964, chapter 7) and corresponding to the latent roots in order of decreasing magnitude, the canonical variates described in sections 6.4 and 6.5 turn out to be

$$X_1 = 5.52x_1 + 3.24x_2 + 2.63x_3, \tag{8.1}$$

$$X_2 = 4.66x_1 - 3.91x_2 + 15.94x_3, \tag{8.2}$$

$$X_3 = -2.40x_1 + 3.25x_2 + 5.43x_3. \tag{8.3}$$

We have indicated in section 6.5 that the canonical variate X_1 corresponding to the largest latent root of (6.5) will show the largest variation between groups, and that any canonical variate that corresponds to a linear relationship between variates will remain constant from group to group within limits that can be accounted for by the random departures. Equation (6.10) provides a test for linear relationships. In the light of these remarks it is of some interest to examine the values of X_1, X_2, X_3 when the annual means of x_1, x_2, x_3 are substituted into (8.1) to (8.3) in turn. These values are given in table 8.2.

TABLE 8.2. Canonical variate values for annual means
of x_1, x_2, x_3 given in table 8.1.

Canonical variate	X_1	X_2	X_3
Year			
1954	18·01	−1·13	11·36
1955	18·09	2·51	10·99
1956	24·79	1·66	10·94
1957	23·62	−0·83	11·03
1958	26·78	3·94	11·35

Clearly X_3 is more nearly constant from year to year than either X_1 or X_2. Application of (6.10) indicates that only X_3 is consistent with a linear relationship, and the form of the relationship is estimated by equating X_3 given by (8.3) to a constant so chosen that the estimated line passes through the mean of all observations. The linear relationship may be written

$$-2.40x_1 + 3.25x_2 + 5.48x_3 = 11.14. \qquad (8.4)$$

Before discussing this equation further we note in passing that X_1 has low values for the first two years and high values for the last three years. As the trees were just carrying their first crops of measurable size in 1954, biologists would probably regard X_1 as measuring an establishment effect. An inspection of the data in table 8.1 indicates that both crop and wood growth show an increasing trend with time, both showing a marked change between 1955 and 1956. This is reflected in X_1 where these two variates are more heavily weighted. X_2 shows a less consistent pattern of variation than X_1.

The same variates for a similar group of trees of the variety Worcester Permain in the same orchard were also subjected to a canonical analysis. Although the X_1 for Cox and Worcester had very different coefficients, the X_1 for Worcester, like that for Cox, had low values for 1954 and 1955, but high values in the later years.

Returning now to (8.4) we note that opposite signs are associated with the cropping variable x_1, and the vegetative variables x_2, x_3, suggesting the balance in the two types of growth previously mentioned.

Equation (8.4) may be rewritten

$$x_1 = 1{\cdot}35x_2 + 2{\cdot}28x_3 - 4{\cdot}64, \tag{8.5}$$

and in this form its interpretation has been discussed in Sprent (1968). If we take anti-logarithms the suggestion is that crop is proportional to a power of wood growth slightly in excess of unity and to a power of girth increment slightly in excess of two. The reader who believes in simple laws of nature might well ask whether random departures in the data can account for these coefficients not being one and two exactly. Williams (1959, chapter 11) shows how a p-variate generalization of the method introduced in section 6.3 may be used to test whether a hypothetical function of the form

$$x_1 = \beta_1 x_2 + \beta_2 x_3 + \text{const.}$$

for specified β_1, β_2 is compatible with the data. For the given data the test shows

$$x_1 = x_2 + 2x_3 + \text{const.} \tag{8.6}$$

to be compatible.

Before the experimenter puts too much faith in (8.6) it is worth pointing out that the test also shows the data to be compatible with either of the relationships

$$x_1 = 2x_2 + 3x_3 + \text{const,} \tag{8.7}$$

$$(\sqrt{2})x_1 = 2x_2 + 3x_3 + \text{const.} \tag{8.8}$$

Enthusiasts for simple laws may prefer (8.7) or (8.8) to (8.6), or at least make a case for them as logical alternatives. If the reader feels that the inability to distinguish between hypothetical relationships as different as (8.6), (8.7) and (8.8) makes the method rather useless, we hasten to point out that only a very small batch of data with relatively high variability within years was used for illustrative purposes. so that a reader with only limited computational facilities at his disposal could perform the computations himself. With more data some or all of (8.6), (8.7) and (8.8) may be rejected.

Example 8.2. In this example we consider some hypothetical data, explaining how it was obtained and the nature of the random departures from underlying relationships. In this example, it is the sub-space determined by the linear relationships that is of interest. We show that different ways of determining that sub-space lead to different linear relationships, but that these all determine very much the same sub-space.

The data are given in table 8.3; the same data have been discussed from a slightly different viewpoint in Sprent (1968). They represent hypothetical measurements on units, and were obtained first by determining sets of values that satisfied two given *non*-linear relationships exactly; to the measurements thus obtained, certain small errors were added, mostly less than about 5 per cent of the true value. These were added in a haphazard (as distinct from a truly random) manner.

It is suggested that one or more of the following hypotheses may hold

$$H_1: x_r = k_1 x_3 x_4{}^2, \quad r = 1, 2; k_1 \text{ constant}, \tag{8.9}$$

$$H_2: x_r = k_2 x_3 x_4, \quad r = 1, 2; k_2 \text{ constant}, \tag{8.10}$$

$$H_3: x_r = k_3 x_3 x_4^{-1}, \quad r = 1, 2; k_3 \text{ constant}, \tag{8.11}$$

$$H_1: x_1 \text{ accords with } H_1 \text{ or } H_2 \text{ or } H_3 \text{ and } x_2{}^3 = k_4 x_1{}^2. \tag{8.12}$$

Hypotheses of the type H_1 to H_4 are not unreasonable in certain practical situations. For example, if x_3, x_4 represent height and diameter of a cylindrical container, e.g. a tin of pork and beans, and the pork is uniformly distributed throughout all tins, then if x_1 represents the weight of pork in a tin it follows that x_1, x_3 and x_4 would be related by an equation of the form (8.9) and H_1 would hold. In this example we could be fairly specific about where any departures from H_1 are likely to occur in practice—namely in the amount of pork in a tin of given dimensions; assuming we measure carefully it is a matter of practical experience that tins are normally true cylinders, the volume of which can be determined very accurately. In Sprent (1968) another situation was discussed. It was suggested that x_1 and x_2 might represent the amounts of two chemical constituents, A and B, in a bone, and x_3 the length and x_4 the diameter of the bone. If we are prepared to assume the bone is approximately cylindrical, then apart from departures attributable to non-uniform distribution of chemicals in the bone and failure of the bone to be completely cylindrical, H_1 implies that the amount of chemical is proportional to the volume of the bone, and H_2 that it is proportional to the cylindrical surface area of the bone. On the other hand, H_3 implies that the amount of chemical present is directly proportional to the surface area and inversely proportional to the cross-sectional area of the bone. The hypothesis H_4 would be relevant if the amount of the second chemical, B, depended upon a catalytic reaction involving the first chemical, A, in such a way that the amount produced is proportional to the surface area of the substance A present. In more general contexts H_1 and H_2 are important when we are considering quantities proportional to areas and volumes of cylinders and certain other regular shapes; H_3 is a type of joint proportionality relationship that is often met, and H_4 is an example of a situation where one variate, x_2, depends upon two others x_3, x_4 through an intermediate variate x_1.

An important feature of H_1 to H_4 is that they represent linear relationships between logarithms of the variates. Similar sets of hypotheses may be set up in other fields of study. An economist, for instance, may postulate that people's demand for a commodity may be directly proportional to their income and inversely proportional to the square of the distance of their homes from the nearest supplier. A linear relation between logarithms would again hold for such a model.

As the person who generated the data in table 8.3 the author

TABLE 8.3. Hypothetical measurements on units.

x_1	x_2	x_3	x_4
7·95	2·40	1·03	1·93
4·07	2·24	2·09	1·05
6·17	3·72	3·03	1·00
33·11	8·91	4·27	2·06
14·79	4·47	1·95	2·16
100·00	14·80	2·89	3·89
24·55	7·09	2·90	1·86
8·32	4·90	3·89	1·10
74·13	13·80	4·21	3·12
131·81	19·50	4·07	4·29

knows perfectly well what sort of linear relationships should appear in an analysis, but an experimenter having observed data of this type is usually not in such a fortunate position—if he were he would hardly need to be doing the experiment. He is, however, often in a position, as the situation outlined above suggests, to decide that it is reasonable to assume that certain general types of hypotheses (e.g. those of the type H_1 to H_4) may hold, although he is uncertain precisely which. If these are the only types of hypothesis he can reasonably visualize, then clearly it is worth looking for linear relationships between logarithms of the data. With the data in table 8.3 we cannot proceed as we did in example 8.1, for with these data there is no grouping of any sort to give an estimate of W. It is in situations such as this that many experimenters resort to least-squares regression methods, arbitrarily selecting one variate as the dependent variate and labelling the remainder regressors. We shall have more to say about the pitfalls of this approach in section 8.2. What can be done more usefully, and the assumptions involved, can only be discussed in terms of concrete ideas about what our data represent. For purposes of discussion we therefore follow Sprent (1968) and assume x_1 and x_2 are chemical measurements on bones and x_3 and x_4 are physical measurements of the type described above.

We make no prior assumptions as to which of the hypotheses H_1 to H_4 are the more likely. Writing $y_r = \log x_r$ we shall work with the y_r as our 'observed' variates. As we have no means of estimating W, we must make some assumptions about it.

For any real bone there will be some departures from true cylindrical shape that will reflect itself in errors in the calculated volume based upon the assumption that it is cylindrical. Put another way, x_3 and x_4 will not be true measures of the length and diameter of a cylinder. In practice the experimenter will have to adopt some convention about how he measures these quantities, e.g. he may decide to measure diameter as the mean of two readings at right angles to one another taken half way along the length of the bone, or he may record it as the mean of diameter readings taken at both ends of the bone, together with a reading in the middle. The chemical measurements will also be subject to departures, perhaps on account of difficulties in chemical analysis, but more usually because of sampling variations due to uneven distribution of chemical in the

129

bone. For any one bone there will almost certainly be some correlation between all these departures, but if we expect these departures within a bone to be small compared to the big variations in measurements from bone to bone, it may be hoped that the form of W will not greatly affect the determination of relevant relationships. As a first assumption we shall suppose $\mathbf{W} = \sigma^2\mathbf{I}$. It is clearly stretching credulity to believe that such equality and independence of departures will hold for the logarithms of a set of physical and a set of chemical measurements; use of the approach must depend entirely on the assumption that the precise form of $\mathbf{W}$ is not very important in the presence of only small departures from the exact relationship.

We require then the latent roots and vectors of

$$|\mathbf{B} - \lambda\mathbf{I}| = 0. \tag{8.13}$$

At this stage the precise value of σ^2 is not important. Determination of the latent roots and vectors of (8.13) where $\mathbf{B}$ is the sum of squares and products between units is a little simpler here than in the general determination of canonical variates, and corresponds to the problem of principal component analysis, as discussed by Seal (1964, chapter 6). For the given data the latent roots of (8.13) account respectively for 91·95, 7·89, 0·11 and 0·05 per cent of the variation between units. It seems reasonable to regard the last two as specifying variation due to departures from the model. Methods of testing the reasonableness of this assumption are discussed briefly by Sprent (1968), and clearly they are special cases of the result in (6.10), although more general methods exist for testing equality of roots for the population when (8.13) applies; details are given in Anderson (1963).

The tests indicate that the latent vectors corresponding to the two smallest latent roots may be associated with linear relationships. These roots and vectors may be computed by methods given by Kendall (1957) or by Seal (1964, chapter 6). The latent vectors corresponding to the smallest two roots are given by the coefficients in the relationships

$$-0{\cdot}1046y_1 - 0{\cdot}4593y_2 + 0{\cdot}5485y_3 + 0{\cdot}6908y_4 = k_1, \tag{8.14}$$

and

$$0{\cdot}5665y_1 - 0{\cdot}6803y_2 + 0{\cdot}1091y_3 - 0{\cdot}4534y_4 = k_2. \tag{8.15}$$

The hypotheses H_1, H_2, H_3 all suggest that more meaningful relationships may be obtained by taking linear combinations of (8.14) and (8.15) in such a way that first y_2 and then y_1 is eliminated. Elimination of y_2 leads to

$$y_1 = 0 \cdot 98 y_3 + 2 \cdot 05 y_4 + \text{const.} \tag{8.16}$$

and elimination of y_1 leads to

$$y_2 = 0 \cdot 97 y_3 + 1 \cdot 03 y_4 + \text{const.} \tag{8.17}$$

If the precise value of the constants are of interest, they may be determined by using the fact that the relationships pass through the means of all observations. The coefficients in (8.16) and (8.17) are all close to one or two, and departures from these values could well be ascribed to observational departures from the true relationship.

Note that (8.14) and (8.15) specify the same two-dimensional sub-space of a four-dimensional space as do (8.16) and (8.17), but in terms of the hypotheses, we are interested in the latter pair of relations which are in themselves more meaningful than the former. In passing we also note that (8.14) and (8.15) exhibit the principal component property of orthogonality, i.e. the sums of the products of the coefficients of the y_i are zero, but (8.16) and (8.17) do not possess this property, i.e. they do not represent hyperplanes at right angles to one another.

Other assumptions about **W** are possible. If it could be assumed that y_1 and y_2 were each linear functions of y_3 and y_4 and the latter pair could reasonably be regarded as regressors, an approach via multiple regression in the general sense leading to equation (5.1) becomes appropriate. Strictly this approach is not justifiable if y_3 and y_4 are known to contain random departures, for one would then expect something like a generalization of the attenuation effect described in section 3.1 to occur. If it were felt that the error in assuming the bones to be cylindrical were small, one could at least hope that the regression approach would be sufficiently informative to suggest which hypotheses were appropriate. Note that this approach is not suitable for testing the hypothesis H_4.

Denoting the 4×4 covariance matrix for the logarithmic data by S, S being our estimate of Σ in section 5.1, S may be partitioned into 2×2 matrices.

$$\begin{pmatrix} S_{11} & S_{12} \\ S_{21} & S_{22} \end{pmatrix}$$

and the matrix of estimated regression coefficients occurring in (5.1) is

$$S_{12}S_{22}^{-1}$$

leading to regression equations

$$y_1 = 0 \cdot 96 y_3 + 2 \cdot 04 y_4 + \text{const.}, \qquad (8.18)$$

$$y_2 = 0 \cdot 96 y_3 + 1 \cdot 03 y_4 + \text{const.} \qquad (8.19)$$

in close agreement with (8.16) and (8.17).

Yet another approach is by way of canonical correlations, briefly discussed in section 6.5. We seek linear functions of the chemical variates and linear functions of the physical variates highly correlated with one another. Following Anderson (1958), the canonical variates are obtained by using the latent vectors of (6.11) or (6.12) with Σ replaced by S, i.e. by using

$$|S_{12}S_{22}^{-1}S_{21} - \lambda S_{11}| = 0,$$

and the variates turn out to be

$$X_1 = -2 \cdot 54 y_1 + 1 \cdot 14 y_2 + \text{const.}$$

$$Y_1 = -1 \cdot 47 y_3 - 3 \cdot 96 y_4 + \text{const.},$$

and these have a correlation coefficient of $0 \cdot 99$. The second pair of canonical variates are

$$X_2 = 7 \cdot 09 y_1 - 12 \cdot 22 y_2 + \text{const.}$$

$$Y_2 = -5 \cdot 06 y_3 + 1 \cdot 95 y_4 + \text{const.},$$

and these have a correlation coefficient of $0 \cdot 96$.

Clearly both X_1 and Y_1 and also X_2 and Y_2 are highly correlated. The method of standardizing the latent vectors means that this implies approximate relationships of the form

$$X_1 = Y_1 \qquad \text{and} \qquad X_2 = Y_2.$$

132

Written in terms of the y_i these relationships look quite different from anything obtained so far, but they define a very similar space to (8.16) and (8.17) or any of the other pairs of relationships, for on eliminating first y_2 and then y_1 we get

$$y_1 = 0 \cdot 99 y_3 + 1 \cdot 94 y_4 + \text{const.} \qquad (8.20)$$

$$y_2 = 1 \cdot 02 y_3 + 1 \cdot 02 y_4 + \text{const.} \qquad (8.21)$$

Again, this approach would not be suitable for examining a hypothesis like H_4.

It should be stressed that the above artificial data are almost too good to be true. In many fields where these techniques are used, particularly in biology and economics, departures from relationships are more marked, and the agreement between the three methods would not usually be so good.

8.2. Regression with arbitrary choice of regressors.

In example 8.2 we considered three rather different approaches chosen in the light of hypotheses we believed might hold, and each led to rather similar conclusions. We indicated that past workers faced with observations on several variates, suspecting at least one linear relationship to hold and believing their errors to be small, have decided to select arbitrarily one variate to be the dependent variate and have taken the remaining ones to be regressors. It is to be hoped that the discussion both in earlier chapters and that associated with example 8.2 will make clear the futility of this approach except when there is only *one* relationship and departures from it are small. In this case it would be a particularization of the multiple regression approach used in example 8.2, providing the dependent variate could be picked on some basis that made it logically distinct from the others.

When more than one relationship exists, but only one is estimated by a method such as arbitrarily selecting a dependent variate and regressing on all others, this merely selects one of an infinity of defining hyperplanes. Such an equation may be useful for predicting dependent variate values given further values of the regressors, but for testing the validity of hypotheses it is as useless as if we had equation (8.14), say, but not (8.15).

8.3. How linear relationships may arise in practice

Experimenters are often surprised how well one or more relationships may fit data, but it is sometimes not unreasonable that they should.

In measurements on growing organisms for example, a size measurement may be made for four characteristics at each of a number of discrete time points. Good approximations to the growth curves for each characteristic can often be obtained by polynomials in time, t, of low degree. Suppose polynomials of degree three suffice to specify the growth curves of each of the four characteristics, e.g.

$$y_i = a_i + b_i t + c_i t^2 + d_i t^3, \ i = 1, 2, 3, 4, \tag{8.22}$$

then using (8.22) with $i = 1, 2, 3$ we can express t, t^2, t^3 as linear functions of y_1, y_2 and y_3. Substitution in (8.22) with $i = 4$, then gives a linear relationship between y_1, y_2, y_3 and y_4. In general, if p characteristics all have growth curves expressible approximately by polynomials of degree r, where $r < p$, there will be $p-r$ approximate linear relationships between the variables representing the p measurements.

In section 4.5 we referred to allometric relationships and simple allometry with two variables. The concept may be extended to p variables. Hopkins (1966) suggests that there are good reasons for associating allometry with a straight line in p dimensions. This implies $p-1$ linear relationships holding approximately between the logarithmic variables. Hopkins discusses the subject at some length. Some writers prefer to consider a generalization simply to $p-r$ linear relationships, but the arguments associated with (8.22) suggest that given enough characteristics, and reasonably orthodox growth patterns for each, there are bound to be some near linear relationships between the measured variates, and in view of the general smoothing tendency of the logarithmic transformation for many types of data probably an even greater number of approximately linear relationships between logarithms. Unless we restrict the definition to a straight line in p dimensions we are likely to find ourselves saying that all growth is allometric if we measure enough characteristics! We therefore recommend that the term 'simple allometry' be reserved for a straight line in p dimensions. This represents $p-1$ linear relationships, or as they are sometimes called in this context, constraints. The reader is referred to Hopkins's paper for more details.

8.4. Other methods of estimating functional relationships

All methods discussed so far for estimating functional relationships have implied some knowledge of the departure structure, or at least a fairly strong assumption about it. Estimates of the departure covariance matrix have often been obtained from replication. In example 8.1, for instance, there was replication within years. Thus we grouped our data by making use of a further variable *time*. Such a variable is referred to as an *instrumental* variable. More general types of instrumental variable may be used. In general, we require them to have the property that they are correlated with the mathematical variables of the functional relationship, but not with the departures exhibited by the observed variates in the case of homogeneous error structure independent from observation to observation. Kendall and Stuart (1967, chapter 29) discuss instrumental variables. A special case of the use of grouping in estimating a functional relationship in the bivariate case is due to Wald (1940). If the relationship to be estimated is of the form

$$\eta = \alpha + \beta \xi$$

and the number of observations is even, then we can obtain consistent estimators $\tilde{\alpha}$, $\tilde{\beta}$ of α, β by dividing the observations into two equal groups providing a certain condition given below in (8.25) holds. If this holds, then if $\bar{x}_1$, $\bar{x}_2$, $\bar{y}_1$, $\bar{y}_2$ denote the means of the observed x in the first and second group and the observed y in the first and second group then

$$\tilde{\beta} = \frac{\bar{y}_2 - \bar{y}_1}{\bar{x}_2 - \bar{x}_1} \tag{8.23}$$

and

$$\tilde{\alpha} = \bar{y}_2 + \bar{y}_1 - \tilde{\beta}(\bar{x}_2 + \bar{x}_1). \tag{8.24}$$

The condition for consistency is that

$$\lim_{n \to \infty} \inf |\xi_2 - \xi_1| > 0, \tag{8.25}$$

where ξ_1, ξ_2 are the true variable values, of which $\bar{x}_1$, $\bar{x}_2$ are estimates Theoretically, unless more is known about errors, there is no guarantee that any allocation to groups will ensure that (8.25) holds. However, if the departures are not too big compared to the spread of x values it is almost certain that (8.25) will hold if the $n/2$ smallest values of x are put in one group and the $n/2$ largest are put in the other group. In this case (8.23) and (8.24) yield consistent estimators. If the departures are assumed homogeneous with that in ξ uncorrelated with that in η, it is possible to obtain consistent estimators of $\sigma_{\delta\delta}$ and $\sigma_{\varepsilon\varepsilon}$. Although consistent, these estimators are often not very reliable, it being possible for them to take negative values. An example given by Kendall and Stuart.

Bartlett (1949) proposed dividing the observations into three groups, one group containing the lower values of x, a second group containing moderate values of x, and a third group containing the higher values of x. The slope is estimated by an analogous formula to (8.23), only the means involved are that for the first and third groups. There remains the question of allocation to groups. Kendall and Stuart suggest that work to date indicates that if the x are assumed error free, then for a wide range of symmetric distributions of X we should put one third of the observations in each group, to reach near maximum efficiency. Bartlett has shown that if x is uniformly distributed and this allocation is made, with x assumed error free, the efficiency of $\tilde{\beta}$ is greater than that of the estimate using only two groups.

Exercises

8.1. The following sets of observations are believed to satisfy one or both of the following hypotheses, subject only to small departures. Investigate this possibility and obtain point estimates of the parameters $\alpha, \gamma, \delta, k_1, \beta, k_2$ if the hypotheses appear reasonable:

$H_1: x_1 x_2{}^{\alpha} x_3{}^{\gamma} x_4{}^{\delta} = k_1,$

$H_2: x_3 = k_2 x_4{}^{\beta}.$

x_1	x_2	x_3	x_4
1·0	9·1	5·2	2·8
4·9	3·2	8·1	7·1
6·7	2·9	6·9	5·5
3·8	2·1	24·8	70·2
4·1	3·4	9·3	9·0
11·2	1·7	10·6	13·2
5·1	2·2	15·2	24·7
2·2	4·3	10·7	13·0

8.2. Show that if departures of observations x_i, y_i from ξ_i, η_i are uncorrelated, homogeneous and independent, Wald's method of grouping gives consistent estimators of $\sigma_{\delta\delta}$, $\sigma_{\varepsilon\varepsilon}$.

(Kendall and Stuart, (1967, chapter 29)).

8.3. Using Wald's method of grouping, estimate the functional relationship $\eta = \alpha + \beta\xi$ from the observations x_i, y_i

x_i	0	1	2	3	4	5	6	7	9
y_i	7·2	4·9	3·1	0·9	−1·2	−2·9	−5·3	−7·1	−10·9

Explain how you have dealt with the odd number of observations in assigning to groups. Examine also Bartlett's (1949) method using three groups for the above data.

CHAPTER 9

Miscellaneous Topics in Linear Regression

9.1. The probit regression line

In this chapter we deal with several miscellaneous topics to complete our treatment of the linear case. In earlier chapters we have considered weighted least squares and generalized least squares in some detail. In most cases maximum likelihood solutions were found to be equivalent to least squares or some generalization thereof.

In section 4.5 we considered transformations of the data designed to achieve linearity. The aim of these transformations was essentially to reduce the problem to one of a standard form that had already been investigated. The *probit* transformation we are about to consider is one of several that have been used to transform a sigmoid curve to a straight line; it is appropriate for transforming the sigmoid curve of the normal distribution function. We consider this particular transformation because it is easy to follow how it arises in practice and because it is widely used and has been discussed in considerable detail, and illustrates a number of points of principle. The standard work on this topic is by Finney (1952). Here we merely outline the way in which the model arises and discuss briefly the associated estimation procedures as far as point estimation of the regression parameters. Readers wishing to use the method in practice are recommended to refer to a detailed treatment such as that given by Finney.

The method is often used in biological assays where a so-called quantal response is observed. As a simple example, suppose an insecticide is applied at a known concentration to a batch of insects. If the concentration were very low none may die, and if it were very high all may die. In comparison of different insecticides to

138

determine their relative effectiveness interest attaches to concentrations at which certain percentages of the insects die. Whether or not a particular insect dies when exposed to a given concentration, or dose, d, say, of insecticide depends upon that individual's *tolerance*. There will be a distribution of tolerances over a population of insects, and those with tolerances less than d will die when given a dose of that concentration. In practice the distribution of tolerances is usually rather skew, but if dose is replaced by $x = \log d$, it is then often found that the distribution of tolerances is approximately normal. If the distribution over a population is in fact normal, then the proportion killed for a log (dose) equal to x is given by

$$P_x = \frac{1}{\sqrt{(2\pi)}\sigma} \int_{-\infty}^{x} \exp\left\{-(x-\mu)^2/2\sigma^2\right\} dx. \tag{9.1}$$

The variable x is often called the dosage.

If each individual in a batch of n responds independently, then the probability that exactly r die when a dosage x is applied is

$$p_{rx} = {}^nC_r P_x^r (1 - P_x)^{n-r}. \tag{9.2}$$

The practical problem to be solved is the evaluation of μ, σ in (9.1), given the percentage or proportion killed in batches exposed to various doses. The graph of P_x as a function of x has the well-known sigmoid form, and μ is of interest as the dosage that gives a 50 per cent kill. The corresponding dose, d, is often called the *median effective dose* (ED50) or the *median lethal dose* (LD50). The parameter σ^2 measures the spread of tolerances; the smaller σ^2, the greater the effect upon mortality of any change in dose.

The *probit* of P_x is defined in terms of a normal distribution of mean 5 and variance 1. It is in fact the abscissa corresponding to a cumulative probability P_x for the distribution $N(5, 1)$, and is therefore given by y in

$$P_x = \frac{1}{\sqrt{(2\pi)}} \int_{-\infty}^{y-5} \exp\left(-x^2/2\right) dx. \tag{9.3}$$

139

Comparing (9.1) and (9.3) it is easily seen that the probit of the expected proportion killed in any batch satisfies the equation

$$y = 5 + (x - \mu)/\sigma. \tag{9.4}$$

Thus the logarithm of the median effective dose is the value of x for which $y = 5$. The use of the normal distribution with a mean of 5 in the definition of a probit is simply a ruse to avoid certain negative values that would occur if standard normal deviates were used instead. We may rewrite (9.4) as

$$y = \alpha + \beta x, \tag{9.5}$$

and if α, β are estimated from a set of observed values of x and y it is a simple matter to obtain estimates of μ, σ.

Estimation of α, β may look at first sight like an ordinary least squares regression problem, but we have said nothing about the departure structure. In fact, our basic data consist of observed dosages $x_1, x_2, \ldots, x_n$ and corresponding to a dosage x_i we have a proportion $p_i = r_i/n_i$ killed, where r_i is the number killed out of a total of n_i exposed. From equation (9.2) it is clear that $E(p_i) = P_i$, where P_i is the value of P_x in (9.1) when $x = x_i$.

Using (9.2) the likelihood function for the observed proportions killed is

$$L = \sum_{i=1}^{n} \frac{n_i!}{n_i! \, (n_i - r_i)!} \, P_i{}^{r_i} (1 - P_i)^{n_i - r_i}.$$

Taking logarithms, $L^* = \ln L$ is

$$L^* = \sum r_i \ln P_i + \sum (n_i - r_i) \ln (1 - P_i) + \text{const.}$$

In itself P_i is not of particular interest as we only require it indirectly because it is a function of α, β. Differentiation of L^* with respect to α, β and equating the derivatives to zero gives the normal equations. Writing $\theta = \alpha, \beta$ in turn the normal equations are

$$\frac{\partial L^*}{\partial \theta} = 0,$$

i.e.

$$\sum \frac{r_i}{P_i} \frac{\partial P_i}{\partial \theta} - \sum \frac{n_i - r_i}{1 - P_i} \frac{\partial P_i}{\partial \theta} = 0,$$

or

$$\sum \frac{n_i(P_i - p_i)}{P_i(1 - P_i)} \frac{\partial P_i}{\partial \theta} = 0. \tag{9.6}$$

An analytic solution of (9.6) is not possible. The procedure used is to start with estimates α', β' of α, β and to adjust these by an iterative process. The initial estimates α', β' are usually obtained graphically. Adjustments $\delta\alpha$, $\delta\beta$ may be obtained from Maclaurin's expansion to the first order of small quantities and are given by

$$\frac{\partial L^*}{\partial \alpha'} + \delta\alpha \frac{\partial^2 L^*}{\partial \alpha'^2} + \delta\beta \frac{\partial^2 L^*}{\partial \alpha' \partial \beta'} = 0, \tag{9.7}$$

$$\frac{\partial L^*}{\partial \beta'} + \delta\alpha \frac{\partial^2 L^*}{\partial \alpha' \partial \beta'} + \delta\beta \frac{\partial^2 L^*}{\partial \beta'^2} = 0, \tag{9.8}$$

where the primes on α, β imply the value of the derivatives at α', β'. The second-order derivatives may be simplified by replacing empirical values p_i by P_i after differentiating. In this way (9.7) and (9.8) reduce to

$$\delta\alpha \sum \frac{n_i}{P_i Q_i}\left(\frac{\partial P_i}{\partial \alpha'}\right)^2 + \delta\beta \sum \frac{n_i}{P_i Q_i} \frac{\partial P_i}{\partial \alpha'} \frac{\partial P_i}{\partial \beta'} = \sum \frac{n_i(p_i - P_i)}{P_i Q_i} \frac{\partial P_i}{\partial \alpha'} \tag{9.9}$$

$$\delta\alpha \sum \frac{n_i}{P_i Q_i} \frac{\partial P_i}{\partial \alpha'} \frac{\partial P_i}{\partial \beta'} + \delta\beta \sum \frac{n_i}{P_i Q_i}\left(\frac{\partial P_i}{\partial \beta'}\right)^2 = \sum \frac{n_i(p_i - P_i)}{P_i Q_i} \frac{\partial P_i}{\partial \beta'} \tag{9.10}$$

where $Q_i = 1 - P_i$.

Equations (9.9) and (9.10) may be solved for $\delta\alpha$, $\delta\beta$ and an improved solution is obtained by putting $\alpha'' = \alpha' + \delta\alpha$ and $\beta'' = \beta' + \delta\beta$, the iterative process being repeated until the required degree of accuracy is obtained.

From (9.3) it follows that

$$\frac{\partial P_i}{\partial y_i} = \frac{1}{\sqrt{(2\pi)}} \exp\{-\tfrac{1}{2}(y_i - 5)^2\} = z_i$$

where z_i is the ordinate of the standard normal curve at the abscissa $y_i - 5$. Using (9.5) it follows that

$$\frac{\partial P_i}{\partial \alpha} = \frac{\partial P_i}{\partial y_i} \frac{\partial y_i}{\partial \alpha} = z_i, \quad \frac{\partial P_i}{\partial \beta} = \frac{\partial P_i}{\partial y_i} \frac{\partial y_i}{\partial \beta} = z_i x_i.$$

Thus, if α', β' are first approximations, adjustments are given, using (9.9) and (9.10), as solutions of

$$\delta\alpha \sum \frac{n_i z_i^2}{P_i Q_i} + \delta\beta \sum \frac{n_i z_i^2 x_i}{P_i Q_i} = \sum \frac{n_i z_i^2}{P_i Q_i} \frac{p_i - P_i}{z_i} \qquad (9.11)$$

$$\delta\alpha \sum \frac{n_i z_i^2 x_i}{P_i Q_i} + \delta\beta \sum \frac{n_i z_i^2 x_i^2}{P_i Q_i} = \sum \frac{n_i z_i^2 x_i}{P_i Q_i} \frac{p_i - P_i}{z_i}. \qquad (9.12)$$

Writing $w_i = z_i^2 / P_i Q_i$, (9.11) and (9.12) become

$$\delta\alpha \sum n_i w_i + \delta\beta \sum n_i w_i x_i = \sum n_i w_i \frac{p_i - P_i}{z_i}, \qquad (9.13)$$

$$\delta\alpha \sum n_i w_i x_i + \delta\beta \sum n_i w_i x_i^2 = \sum n_i w_i x_i \frac{p_i - P_i}{z_i} \qquad (9.14)$$

where z_i, P_i, Q_i and hence w_i can be computed for given α', β', and appropriate tables are available to aid this. The interesting feature of (9.13) and (9.14) is that they are of the form of weighted least-squares regression equations for the regression of a variate $(p-P)/z$ on x.

We shall not discuss the use and interpretation of probit analysis. Nor do we discuss the various techniques available to reduce computational labour. All these matters are dealt with very adequately by Finney, who also deals with many practical considerations that may lead to modification of the model developed here. For example, the responses of individual insects in a batch may not be independent, or allowance may have to be made for natural mortality during the experiment. In dealing with non-linear regression in chapter 10 we shall see that iterative solutions are there the rule rather than the exception.

An alternative transformation based on a logistic distribution is known as the logit transformation. It is discussed by many writers including Berkson (1953).

9.2. Berkson's model with controlled variables.

The second miscellaneous topic we deal with is the so-called controlled variable model introduced by Berkson (1950). It is of practical importance because it represents a situation of fairly common occurrence, and it is of theoretical interest as it indicates the care needed to define clearly a model.

In the physical sciences, in determining a law-like relationship, it is common practice to 'set' the value of one of the measured quantities at pre-selected fixed values. For example, the current passing through a circuit may be set at 1 amp (or more generally at x_1 amp) on a meter, and a reading y_1 taken of some response of interest, e.g. potential difference. The circuit is then adjusted so that the meter reads 2 amp (x_2 amp in general) and the response y_2 recorded, and so on.

It may well happen that the meter reading is subject to random errors, and instead of a current x_i amp as recorded by the meter at the ith reading actually flowing in the circuit, the true current has an unknown value ξ_i, where $\xi_i = x_i + \delta_i$.

Assuming the usual model for y_i, i.e. that its distribution is normal with some appropriate mean and variance σ^2, we have a model that looks superficially like those discussed in chapter 3. However, there is a difference. In that chapter the situation was such that in repeated experimentation, if we had the same fixed (but unknown) ξ_i we would observe different values of x_i. However, if we are to repeat the experiment considered here with x_i fixed, each fixed x_i will correspond to a different true ξ_i. In fact, x_i is now a fixed (mathematical) variable and ξ_i is a (random) variate, the random component being δ_i. If we assume that for all i, δ_i is $N(0, \sigma_\delta^2)$, together wih the usual independence condition, our problem is to estimate α, β in the model

$$y_i = \alpha + \beta \xi_i + \varepsilon_i.$$

If we make the classical regression assumptions about the distribution of ε_i, replacing ξ_i by the known x_i we get

$$y_i = \alpha + \beta x_i + \varepsilon_i - \beta \delta_i. \tag{9.15}$$

F

This differs notationally from a similar model arising with functional relationships in so far as x_i is now a (mathematical) variable rather than a (random) variate, and it is not correlated with δ_i. Thus the independence of error upon x is retained and classical regression methods of estimation of α, β from (x_i, y_i) are appropriate despite the appearance of (9.15).

9.3. Bias due to ignoring correlations between departure

One of the most striking features of classical least-squares regression as opposed to the more general methods of sections 4.1 and 5.6 is that in the classical case no knowledge of σ^2 is required for point estimation of β, and as a consequence of this an estimate of σ^2 can be obtained using the residual sum of squares about the fitted line.

On the other hand, when the departures, ε_i, may be correlated, it is necessary even for point estimation of β to have at least an estimate of the departure covariance matrix $\mathbf{\Sigma}$. Without replication of y values for fixed x this is almost impossible, and one may have to resort to guesses. In the more complicated problems of law-like relationships dealt with in section 8.1 we showed that progress could only be made in the absence of knowledge about $\mathbf{\Sigma}$ if we were prepared to make simple assumptions about it.

In the regression problem with correlated departures, Watson (1955) has studied the effect of a wrong choice of $\mathbf{\Sigma}$. In a more recent paper, Watson (1967), an effort has been made to unify the whole approach to this problem using concepts of spectral analysis and an attempt has been made to bring together the approaches used in analysis of variances and those used in time series. In view of the fact that we have specifically omitted the treatment of time series we shall not go into details here.

Classical regression theory is concerned with the case $\mathbf{\Sigma} = \sigma^2 \mathbf{I}$. Although we have not used such an approach in this book, it can be shown that for a *known* $\mathbf{\Sigma}$ it is always possible to make a change of variables so that for the new variables the covariance matrix has the form $\sigma^2 \mathbf{I}$. Details are given in the earlier of the two papers by Watson. Thus there is no loss of generality in studying the effect of a wrong choice of $\mathbf{\Sigma}$ if we assume the wrong choice is $\mathbf{\lambda} = \sigma^2 \mathbf{I}$. With this choice we would proceed to use classical least square instead of a procedure based upon the true value $\mathbf{\Sigma}$. We shall

144

assume there are n sets of observations and q regressors, the model expressed in the matrix notation of chapter 5 being

$$\mathbf{y} = \mathbf{X\beta} + \mathbf{\varepsilon}.$$

If we assume $\mathbf{\varepsilon}$ is $N(\mathbf{0},\ \sigma^2\mathbf{I})$ when it is really $N(\mathbf{0},\ \mathbf{\Sigma})$ we would proceed by classical least squares to estimate $\mathbf{\beta}$ by

$$\hat{\mathbf{\beta}} = (\mathbf{X'X})^{-1}\mathbf{X'y} \tag{9.16}$$

as in (5.5). We shall study the resultant bias in the estimate of the covariance matrix of the coefficient estimates. This is just one topic considered by Watson (1955) who goes on to examine the effect on F and t tests.

The covariance matrix of $\hat{\mathbf{\beta}}$ in (9.16) is actually

$$(\mathbf{X'X})^{-1}\mathbf{X'}\mathbf{\Sigma}\mathbf{X}(\mathbf{X'X})^{-1} \tag{9.17}$$

but under the classical least-squares assumption it would be taken to be $\sigma^2(\mathbf{X'X})^{-1}$ (see section 5.2 and exercise 5.3), where σ^2 would be estimated by

$$s^2 = \frac{(\mathbf{y} - \mathbf{X}\hat{\mathbf{\beta}})'(\mathbf{y} - \mathbf{X}\hat{\mathbf{\beta}})}{n-q}. \tag{9.18}$$

We may evaluate $E(s^2)$ by noting that

$$\begin{aligned} \mathbf{y} - \mathbf{X}\hat{\mathbf{\beta}} &= \mathbf{y} - \mathbf{X}(\mathbf{X'X})^{-1}\mathbf{X'}(\mathbf{X\beta} + \mathbf{\varepsilon}) \\ &= \mathbf{y} - \mathbf{X\beta} - \mathbf{X}(\mathbf{X'X})^{-1}\mathbf{X'\varepsilon} \\ &= \mathbf{\varepsilon} - \mathbf{X}(\mathbf{X'X})^{-1}\mathbf{X'\varepsilon}, \end{aligned}$$

whence, from (9.18)

$$\begin{aligned} E(s^2) &= \frac{1}{n-q}\ E(\mathbf{\varepsilon} - \mathbf{X}(\mathbf{X'X})^{-1}\mathbf{X'\varepsilon})'(\mathbf{\varepsilon} - \mathbf{X}(\mathbf{X'X})^{-1}\mathbf{X'\varepsilon}) \\[2mm] &= \frac{1}{n-q}\ E(\mathbf{\varepsilon'\varepsilon} - \mathbf{\varepsilon'X}(\mathbf{X'X})^{-1}\mathbf{X'\varepsilon}) \\[2mm] &= \frac{1}{n-q}\ \{\operatorname{tr} \mathbf{\Sigma} - \operatorname{tr}[\mathbf{X'}\mathbf{\Sigma}\mathbf{X}(\mathbf{X'X})^{-1}]\} \tag{9.19} \end{aligned}$$

where $\operatorname{tr}\mathbf{\Sigma}$ is the trace of $\mathbf{\Sigma}$, (i.e. the sum of the principal diagonal elements).

Thus in estimating the covariance matrix of $\hat{\beta}$ as $s^2(\mathbf{X'X})^{-1}$, the bias, noting (9.17), is

$$\mathbf{B} = E\{s^2(\mathbf{X'X})^{-1}\} - (\mathbf{X'X})^{-1}\mathbf{X'}\boldsymbol{\Sigma}\mathbf{X}(\mathbf{X'X})^{-1}.$$

Using (9.19) the bias may be written

$$\mathbf{B} = (\mathbf{X'X})^{-1}\left\{\frac{\text{tr}\boldsymbol{\Sigma} - \text{tr}[\mathbf{X'}\boldsymbol{\Sigma}\mathbf{X}(\mathbf{X'X})^{-1}]}{n-q}\mathbf{I} - \mathbf{X'}\boldsymbol{\Sigma}\mathbf{X}(\mathbf{X'X})^{-1}\right\}. \quad (9.20)$$

In particular, Watson considers the effect of bias on the variance of a single regression coefficient for the special case $\mathbf{X'X} = \mathbf{I}$ implying orthogonality of the $\mathbf{x}$ and a scaling to give what is called an orthonormal set. In this case (9.20) gives for the bias of the estimated variance of $\hat{\beta}_i$

$$B_i = \frac{\text{tr}\boldsymbol{\Sigma} - \sum_{j-1}^{q} \mathbf{x'}_j\boldsymbol{\Sigma}\mathbf{x}_j}{n-q} - \mathbf{x}_i\boldsymbol{\Sigma}\mathbf{x}_j \quad (9.21)$$

where $\mathbf{x}_j$ is the jth column of $\mathbf{X}$.

The extremes of B_i in (9.21) give the maximum and minimum bias. A well known result in matrix algebra states that the extrema of a quadratic form $\mathbf{x'Ax}$, where $\mathbf{x}$ is a unit vector in a subspace spanned by a subset of latent vectors of $\mathbf{A}$ are the greatest and least latent roots of $\mathbf{A}$ associated with that subspace. The lower bound of (9.21) is obtained by *choosing* one of the vectors, say $\mathbf{x}_i$, to be the latent vector of $\boldsymbol{\Sigma}$ corresponding to its largest latent root, and by choosing the remaining $\mathbf{x}_j, j \neq i$ as the latent vectors corresponding to the next $q-1$ largest roots. The upper bound is found by a similar choice with the smallest latent roots. Since $\text{tr}\boldsymbol{\Sigma}$ equals the sum of the latent roots of $\boldsymbol{\Sigma}$ it follows that

$$\text{Sup}(B_i) = \text{mean of } n-q \text{ greatest roots of } \boldsymbol{\Sigma} - \text{least root of } \boldsymbol{\Sigma}$$

and

$$\text{Inf}(B_i) = \text{mean of } n-q \text{ least roots of } \boldsymbol{\Sigma} - \text{greatest root of } \boldsymbol{\Sigma}.$$

Note that unless $\boldsymbol{\Sigma} = \sigma^2\mathbf{I}$, when all latent roots are equal, Sup (B_i) is necessarily positive and Inf (B_i) is necessarily negative. Watson indicates that the generally held belief that correlations tend to make variance estimates deceptively small if they are ignored, seems to be the stronger tendency. Watson's paper should be consulted for details of the effect upon t and F tests of ignoring correlations. Swindel (1968) has obtained bounds without the restriction $X'X = I$.

9.4. Some topics omitted

In section 1.1 we indicated that there was a vast and growing literature on regression and related topics. Some of these have been ignored in this survey, others have received scant treatment. Time series is a topic of great importance, yet for reasons given in section 4.4 we have not treated the subject in detail. Indeed, economic applications and models have been given little attention. For more detail of these the reader could do little better than consult Malinvaud (1966) or Fisk (1966).

Little attention has been paid to simultaneous regression equations. Again, their use has been more common in economics than in other fields. Williams (1959) devotes a chapter to them, and they are also discussed by Fisk in his book.

We have said little about optimum allocation of points, nor have we said anything about models where there are restrictions on the range of permissible values for the regressors, or restrictions on permissible values of the coefficients. Quite recently there have been a number of papers, for example, on determination of confidence bands in linear regression when there are constraints upon the values assumed by regressors. Relevant papers on this topic include those by Bowden and Graybill (1966), Halperin and Gurian (1968) and Dunn (1968).

Perhaps the most severe limitation we have placed upon our models so far has been the requirement of linearity. Once again a comprehensive treatment of non-linearity would demand a book in its own right. However, as it is so important and has not perhaps received the attention it deserves from writers on regression we append a brief introductory chapter on this topic.

Exercises

9.1. Obtain confidence limits for log ED50 estimated by a probit analysis from experimental data. (Finney, (1952)).

9.2. Discuss the problem of testing two or more probit regression lines for parallelism. (Finney, (1952)).

9.3. Examine the bias in variance estimation due to the wrong choice of Σ in estimation of regression coefficient s without the restriction $X'X = I$ that was imposed in section 9.3. (Swindel, (1968))

9.4. Examine the effect on classical t and F tests of ignoring correlations in errors or using a wrong value of Σ. (Watson, (1955)).

CHAPTER 10

Non-linear Models

10.1. Non-linear least squares

In section 5.1 we pointed out that when we said a model was linear we meant that it was linear in its coefficients. Linear models have been shown to include a wide variety of curves and surfaces, e.g. polynomials or surfaces with equations of the form

$$y = \beta_0 + \beta_1 x_1 + \beta_2 x_1 x_2 + \beta_3 x_1 x_2 x_3,$$

say.

There are situations where theoretical considerations may suggest a model that is non-linear in its coefficients or where practical considerations may exclude the possibility of obtaining a reasonable fit with a polynomial. One situation where polynomials are unsatisfactory is that in which there are indications of a distinct asymptote. This situation often arises in studying the response of a crop to a fertilizer. There is generally a maximum amount of fertilizer to which a crop will respond by showing an increase in yield. After a certain level of application, no increase in yield is obtained by adding more fertilizer; there may even be a decrease in yield due to toxic effects, but we shall ignore this complication. While a polynomial may give a close approximation to the fertilizer-yield relationship over the 'active' range of applications, clearly as x, the amount of fertilizer, is increased, y, the yield, if it always accorded with the polynomial, would either increase to infinity or decrease to minus infinity—information not very useful to somebody who wants to know the maximum yield he can expect and the lowest level of fertilizer that will give him something very close to that maximum yield.

Asymptotic curves are in general non-linear in their parameters. We shall see below that the principle of least squares may still be appropriate for estimating parameters, but the normal equations in general will no longer have explicit analytic solutions.

We suppose that the equation to be fitted involves q regressors and k parameters. It may be written

$$y = f(x_1, x_2, \ldots, x_q, \theta_1, \theta_2, \ldots, \theta_k),$$

or in matrix notation

$$y = f(\mathbf{x}, \boldsymbol{\theta}), \tag{10.1}$$

where $\boldsymbol{\theta}$ is the matrix of parameters to be estimated on the basis of n observations satisfying a model

$$\begin{aligned} y_i &= f(x_{1i}, x_{2i}, \ldots, x_{qi}, \theta_1, \theta_2, \ldots, \theta_k) + \varepsilon_i \\ &= f(\mathbf{x}_i, \boldsymbol{\theta}) + \varepsilon_i. \end{aligned}$$

Under the classical regression assumption that the ε_i are independent and $N(0, \sigma^2)$ the likelihood function may be written

$$L = \frac{1}{(\sqrt{(2\pi)}\sigma)^n} \prod_i e^{-\Sigma\{y_i - f(\mathbf{x}_i, \boldsymbol{\theta})\}^2/2\sigma^2}.$$

On taking logarithms it is easily seen that the maximum of L occurs when $\boldsymbol{\theta}$ is chosen to minimize

$$R = \sum\{y_i - f(\mathbf{x}_i, \boldsymbol{\theta})\}^2, \tag{10.2}$$

so that the problem reduces to a least-squares form and the normal equations are

$$\sum_i \{y_i - f(\mathbf{x}_i, \boldsymbol{\theta})\} \frac{\partial f(\mathbf{x}_i, \boldsymbol{\theta})}{\partial \theta_r} = 0, \tag{10.3}$$

$$(r = 1, 2, \ldots, k).$$

If $f(\mathbf{x}, \boldsymbol{\theta})$ is linear in $\boldsymbol{\theta}$ the partial derivatives are functions of x_i only and not of $\boldsymbol{\theta}$. Thus in the linear case the normal equations are a set of k linear equations in k unknowns, and these are immediately soluble if the matrix of coefficients is non-singular. In the general case the equations (10.3) have no explicit solutions and must be solved by iterative methods, using devices such as the Maclaurin expansion in much the way we did for probit regression in section 9.1. Other approaches are possible. Rather than work with normal equations directly it may be preferable to apply the method of steepest descent to equation (10.2); this method is described by Davies (1954). With a high-speed computer, and in particular with only one or two parameters, it may be better to determine the values of R over a grid, gradually taking a finer grid as the minimum is pinpointed, after the manner suggested for finding the minimum of U in section 4.2.

The geometry of non-linear least squares, as well as that of linear least squares, is discussed in some detail by Draper and Smith (1966, chapter 10.)

The solution of (10.3) is not trivial even in very simple cases. Suppose $f(x, \theta) = \theta^x$, i.e. we wish to estimate θ in

$$y = \theta^x \tag{10.4}$$

given n observations (x_i, y_i) and the classical least squares regression assumptions about ε_i hold. From (10.3)

$$\sum_i [(y_i - \theta^{x_i}) x_i \theta^{x_i - 1}] = 0,$$

or

$$\sum_i (y_i x_i \theta^{x_i - 1}) - \sum (x_i \theta^{2x_i - 1}) = 0,$$

and there is in general no explicit solution in terms of the x_i.

It is worth noting that (10.4) implies

$$\log y = x \log \theta \tag{10.5}$$

and θ may be estimated by considering the regression of $\log y$ on x. However, if the classical regression assumptions hold for the model (10.4) they will not hold for (10.5), a point already made in section 4.5. An important case where a transformation to linearity does not exist occurs when the equation to be estimated is

$$y = \alpha + \beta \rho^x \tag{10.6}$$

where α, β and ρ are the parameters.

This is sometimes referred to as the exponential regression and there are several different ways of writing (10.6) that are essentially the same in so far as the parameters in one form can be expressed in terms of the parameters in the other forms. For instance, (10.6) written in the form

$$y = a(1 - br^x) \tag{10.7}$$

or

$$y = p(1 - e^{-q(x + s)}) \tag{10.8}$$

have their parameters related as follows:

$$a = \alpha, \ ab = -\beta, \ r = \rho, \ p = a, \ pe^{-qs} = b, \ \ln r = -q.$$

The precise form in which the exponential regression equation, or Mitscherlich's equation, as it is also called, is written depends on the use to which it is put. In fertilizer experiments where x represents the amount applied and y the yield, the parameters in (10.8) have simple interpretations, e.g. p represents the maximum yield obtainable, s represents a measure of the amount of nutrient available before any fertilizer is added, and q can be regarded as a measure of the efficiency of the fertilizer.

There is a considerable literature on least squares and other methods of estimating the parameters. For this purpose the forms (10.7) or (10.6) have some advantages over (10.8).

For equally spaced values of x, solutions of the appropriate normal equations are simplified, and there are some useful methods that approximate to the least-squares solution. Stevens (1951) gave an ingenious method for estimating the parameters by maximum likelihood. The practical application of that method has been fully explored by Hiorns (1965) who provides helpful tables, extending those given by Stevens. Hiorns claims that the method is easily applied by using a desk calculator, the time required being only fractionally longer than that required to punch a deck of cards.

Approximate methods that may be adequate have been given by Gomes (1953) and Patterson (1956), among others. Recently, Patterson (1969) has discussed the fitting of an equation known as Baule's equation which is a two-regressor analogue of (10.7) involving five parameters, viz.

$$y = a(1 - br^{x_1})\,(1 - cs^{x_2}).$$

In sections 10.2 and 10.3 we discuss two special cases of non-linear regression. That in section 10.2 involves conversion of the problem into a linear weighted least squares set-up. The example in section 10.3 indicates the use of iterative procedures for solving the normal equations (10.3).

10.2. Estimation of parameters in a rational function

Suppose we wish to fit a function of the form

$$y = \frac{f(x)}{g(x)},$$

where $f(x)$, $g(x)$ are polynomials of specified degree p, q respectively.

Turner, Monroe and Lucas (1961) considered this problem given n observations (x_i, y_i) such that

$$y_i = \frac{f(x_i)}{g(x_i)} + \varepsilon_i,$$

where ε_i are independent $N(0, \sigma^2)$. Let $f(x) = \sum_{r=0}^{p} \alpha_r x^r$ and $g(x) = 1 + \sum_{s=1}^{q} \beta_s x^s$. As long as $g(x)$ contains a non-zero constant term there is no loss in writing it in this form, for if $\beta_0 \neq 1$ it can be made so by the reparameterization arising from multiplying numerator and denominator by $1/\beta_0$. If we write

$$y_i = \frac{f(x_i)}{1 + \beta_1 x_i + \beta_2 x_i^2 + \ldots + \beta_q x_i^q} + \varepsilon_i$$

it easily follows that

$$y_i = f(x_i) - \beta_1 x_i y_i - \beta_2 x_i^2 y_i - \ldots - \beta_q x_i^q y_i + g(x_i)\varepsilon_i. \quad (10.9)$$

The problem has now been reduced to fitting a weighted linear regression for y on x, x^2, ..., x^p, xy, $x^2 y$, ..., $x^q y$ using weights $w_i = \{g(x_i)\}^{-2}$. At first sight it is rather alarming to see the dependent variate appearing on both sides of (10.9), and the price we pay for this arrangement to obtain linearity is the appearance of a weighting function that depends not only on x_i but on the parameters $\beta_1, \beta_2, \ldots, \beta_q$, and these are unknown. Regression with unknown weights has been discussed by Williams (1959, chapter 4) and it is clearly a situation where an iterative process is reasonable. Turner, Monroe and Lucas suggest that preliminary estimates of $\beta_1, \beta_2, \ldots, \beta_q$ be chosen and provisional weights be computed as $\{g_0(x_i)\}^{-2}$ where

$$g_0(x) = 1 + \sum \beta_r^0 x^r,$$

β_r^0 being the preliminary estimate of β_r. In many cases it is appropriate to take all $\beta_r^0 = 0$ as a first estimate. This implies doing a preliminary fitting by classical least squares; this will give first estimates

$$\alpha_0', \alpha_1', \alpha_2', \ldots, \alpha_p', \beta_1', \beta_2', \ldots, \beta_p'$$

of the regression coefficients. The β_i' can be used to compute new

153

estimates of the weights $\{g_1(x_i)\}^{-2}$, say, and these provide the basis for a weighted least-squares procedure to obtain new estimates,

$$\alpha_0'', \alpha_1'', \alpha_2'', \ldots, \alpha_p'', \beta_1'', \beta_2'', \ldots, \beta_q''$$

of the coefficients, the process being repeated until, to the required degree of accuracy, there is no alteration in the coefficients in successive iterations. It is not necessary to compute estimates of the α_i at each stage; it is, however, advisable to compute them at successive cycles of the iteration that give stability in the β_j, as a small difference in weights that does not affect the estimates of the β_j may not give stable α_i; if it is found that they are still altering, it may be necessary to iterate further. To obtain new weights, more significant figures will have to be retained in the estimates of β_j to detect changes from the previous cycle. With a large number of coefficients round-off errors may plague attempts to attain too high a precision. The simplest application of the above is to the fitting of a rectangular hyperbola of the form

$$y = \frac{a}{x-b},$$

which can be rewritten

$$y = \frac{\alpha}{1+\beta x} \qquad (10.10)$$

where $\beta = -1/b$ and $\alpha = -a/b$, and estimation of α and β can then proceed as above. If β' is an estimate of β at any stage the estimated weights are $w_i' = (1+\beta' x_i)^{-2}$ and the regression to be fitted is a weighted regression of y on yx. In this example it is easy to get a better starting value than $\beta^0 = 0$, for $x = -1/\beta$ is an asymptote of (10.10) and a rough plot of the points can be used to estimate this. Turner, Monroe and Lucas point out that the final iteration can be used to provide weighted least-squares estimates of σ^2, and that asymptotic standard errors and confidence limits can be obtained for each coefficient in the usual manner for weighted least squares, (see exercise 4.1).

10.3. Fitting the generalized logistic curve

Many empirical growth curves of size against time assume the form of a sigmoid curve. In simple cases these may be symmetrical, but in other cases the 'tail' is longer in one direction than in the other. For the symmetrical sigmoid form a curve that has considerable popularity with biologists is the so-called logistic curve. This is a different form of sigmoid to the normal sigmoid discussed in section 9.1. It is also referred to as the autocatalytic curve.

A logistic curve arises if w represents size and x time when the growth rate dw/dx is proportional to the size already achieved and the amount by which the size falls short of its maximum, i.e.

$$\frac{dw}{dx} = \beta w(\alpha' - w)/\alpha', \tag{10.11}$$

where α' represents the final size achieved, and β is a constant of proportionality. Integration of (10.11) gives

$$w = \frac{\alpha'}{1 + e^{-(\lambda + \beta x)}} \tag{10.12}$$

where λ is a constant of integration dependent essentially on the choice of time origin. If we now write $y = \ln w$ and $\alpha = \ln \alpha'$, (10.12) gives on taking logarithms

$$y = \alpha - \ln\{1 + e^{-(\lambda + \beta x)}\}. \tag{10.13}$$

Nelder (1961) has considered the estimation of the parameters in this equation given n sets of observations (x_i, y_i) under the assumption that

$$y_i = \alpha - \ln\{1 + e^{-(\lambda + \beta x_i)}\} + \varepsilon_i$$

where the ε_i are independent $N(0, \sigma^2)$. Nelder reports that for growing plants the assumption that $y = \ln w$ has constant variance is usually found to be a good approximation. Independence will be guaranteed if different plants are taken at each time x_i. Writing $f(x, \alpha, \beta, \lambda)$, or more briefly $f(x)$ for the right-hand side of (10.13) the least-squares normal equations are given by

$$\sum\{y_i - f(x_i)\} \frac{\partial f(x_i)}{\partial \phi} = 0,$$

155

where $\partial f(x_i)/\partial \phi$ denotes the value of $\partial f(x)/\partial \phi$ at $x = x_i$, and $\phi = \alpha$, β or λ.

Now

$$\frac{\partial f(x_i)}{\partial \alpha} = 1, \quad \frac{\partial f(x_i)}{\partial \lambda} = \frac{e^{-(\lambda + \beta x_i)}}{1 + e^{-(\lambda + \beta x_i)}} = \frac{1}{1 + e^{\lambda + \beta x_i}} = \xi_i, \text{ say,}$$

and

$$\frac{\partial f(x_i)}{\partial \beta} = \frac{x_i e^{-(\lambda + \beta x_i)}}{1 + e^{-(\lambda + \beta x_i)}} = x_i \xi_i.$$

The normal equations have no explicit solutions and iterative procedures are required. Use of Maclaurin's theorem and the replacement of y_i by an empirical value $f(x_i)$ in a similar manner to that employed in the discussion of probits leading to (9.9) and (9.10) in section 9.1, leads to the following adjustment equations once initial values have been obtained by graphical or other means. The adjusting equations are

$$n\delta\alpha + \sum \xi_i \delta\lambda + \sum \xi_i x_i \delta\beta = \sum \{y_i - f(x_i)\} \tag{10.14}$$

$$\sum \xi_i \delta\alpha + \sum \xi_i^2 \delta\lambda + \sum \xi_i^2 x_i \delta\beta = \sum \xi_i \{y_i - f(x_i)\} \tag{10.15}$$

$$\sum \xi_i x_i \delta\alpha + \sum \xi_i^2 x_i \delta\lambda + \sum \xi_i^2 x_i^2 \delta\beta = \sum \xi_i x_i \{y_i - f(x_i)\}. \tag{10.16}$$

The equations (10.14) to (10.16) are of the form of least-squares equations for the regression of $y - f(x)$ on ξ and ξx. Given approximations α', λ', β', these equations may be used to form new approximations $\alpha'' = \alpha' + \delta\alpha'$, $\lambda'' = \lambda' + \delta\lambda'$, $\beta'' = \beta' + \delta\beta'$, where $\delta\alpha'$, $\delta\lambda'$, $\delta\beta'$ are the solutions to (10.14) to (10.16). The iterative cycle is repeated to give estimators $\hat{\alpha}$, $\hat{\beta}$, $\hat{\lambda}$ of sufficient accuracy.

Nelder provides tables to facilitate the computation of the ξ_i. These are based on Berkson's (1953) tables of anti-logits. If $\tau_i = \lambda + \beta x_i$, then $\xi_i = \text{antilogit}(-\tau_i) = 1 - \text{antilogit}(\tau_i)$.

Nelder suggests that convergence may be improved by not making the approximation $y_i = f(x_i)$ after differentiation in the procedures leading to (10.14) to (10.16). The dispersion matrix of the parameter estimates can be obtained by inverting the matrix of coefficients of $\delta\alpha$, $\delta\lambda$, $\delta\beta$ in (10.14) to (10.16) and then multiplying it by an estimate of σ^2. This is a particular example of a procedure with wide applicability in estimation problems, and the matrix that is inverted is known as the information matrix.

Nelder suggests that it may be profitable to use a more sophisticated approach to obtaining starting values than the completely graphical one. He gives a worked example and also discusses the choice of sample points. Clearly, as α is a reflection of the final size achieved, most information on that parameter is given by large values of x, but these values give very little information on the other parameters. Very briefly, Nelder's general conclusion is that for good estimation of parameters the points should cover a very wide range of y values. He also points out that tests of prior hypotheses about parameters will usually need very extensive data for useful results to be attained. General discussions on the allocation of sample points for efficient estimation with non-linear models are given by Chernoff (1953) and Box and Lucas (1959).

The symmetric form of the logistic curve makes it unsuitable for some growth studies and Richards (1959) proposed a generalization which includes other well known curves such as the exponential and Gompertz curves as particular cases. The reader is referred to Richards's paper for a full descriptive and non-statistical discussion of the family, which is derived from the differential equation

$$\frac{\mathrm{d}w}{\mathrm{d}x} = \beta w\{1 - (w/\alpha')^{1/\theta}\}. \tag{10.17}$$

This reduces to (10.11) when $\theta = 1$. For positive θ, Nelder (1961) has shown that on integrating (10.17) we get

$$w = \frac{\alpha'}{\{1 + \mathrm{e}^{-(\lambda + \beta x)/\theta}\}^{\theta}}. \tag{10.18}$$

Taking logarithms and putting $y = \ln w$ and $\alpha = \ln \alpha'$ gives

$$y = \alpha - \theta \ln\{1 + \mathrm{e}^{-(\lambda + \beta x)/\theta}\}. \tag{10.19}$$

Under the same assumptions about departures as we made for the logistic function, Nelder shows that if we write $\rho_i = (\lambda + \beta x_i)/\theta$ and $\xi_i = 1/(1 + \mathrm{e}^{\rho_i})$, then the derivatives of $f(x)$ with respect to

α, λ, β have the same form as they had for the logistic function in terms of the redefined ξ_i, and

$$\frac{\partial f(x_i)}{\partial \theta} = \ln (1-\xi_i) - \tau_i \xi_i = \gamma_i, \text{ say.}$$

Once again an iterative procedure is required and the adjustments can be computed using the information matrix in an analogous manner to that used in fitting the logistic function. The adjustments are computed by solving equations that are essentially the normal equations for the regression of $y - f(x)$ on ξ, ξx and γ. Nelder provides tables of values of γ for given ρ together with a numerical example. In a later paper, Nelder (1962), the restriction that θ be positive is removed by using a reparameterization of the problem.

10.4. Relaxation of assumptions about departures.

In the non-linear case very little work has been done about procedures when there are departures from the underlying relationship in both variates, or when the ε_i are correlated. This would seem to be a field worthy of attention as non-linear estimation is likely to prove increasingly useful in the study of growing organisms, as well as in economic problems.

Mitchell (1968) has recently discussed correlated errors for the exponential function, and Hey and Hey (1960) discussed the particular problem of fitting a hyperbola with both variates exhibiting departures from the true curve.

10.5. Curve fitting and approximation theory

This book has been largely concerned with fitting a curve of supposedly known form to observational data subject to random departures from the values that would have been obtained if the data fitted the curves exactly. The commonly made assumptions of normality lead naturally in most cases to the principle of least squares, or some extension of it. This same principle has been widely used in a non-statistical type of curve fitting often carried out by mathematicians.

Numerical analysts are often interested in replacing a known function $f(x)$, say, which may have a very complicated form, perhaps even with an infinite number of parameters, by a simpler approximating function $g(x, \boldsymbol{\theta})$, say, where $\boldsymbol{\theta}$ is a vector of parameters to be estimated. A common practice is to estimate $\boldsymbol{\theta}$ by least squares. If the approximation is to hold over an interval $[a, b]$, the least-squares procedure is to choose $\boldsymbol{\theta}$ to minimize

$$I = \int_a^b |f(x) - g(x, \boldsymbol{\theta})|^2 \mathrm{d}x. \qquad (10.20)$$

Least-squares has an intuitive appeal in this case, but rather less justification than in statistical problems of curve fitting based on normal theory and the likelihood principle. There are situations where it is preferable to replace (10.20) by

$$I_p = \int_a^b |f(x) - g(x, \boldsymbol{\theta})|^p \mathrm{d}x$$

where p is a specified positive number.

There are many mathematical problems of approximating to functions where a different approach is appropriate. For instance, in engineering design problems it may well be appropriate to minimize the maximum error of approximation, i.e. to minimize in the range of interest the maximum value of $|f(x) - g(\mathbf{x}, \theta)|$. Such a mode of approximation is called Tchebycheff approximation. Approximation theory has developed along rather different lines to statistical curve fitting. An introductory account of linear approximation theory is given by Rice (1964), and of general approximation theory by Davis (1963).

Exercises

10.1. Discuss the estimation of parameters in Mitscherlich's equation. (Stevens (1951), Gomes (1953), Patterson (1956, 1960), Hiorns (1965)).

10.2. Examine the modifications necessary to the procedure developed in section 10.2 if for $g(x)$, $\beta_0 = 0$, and in consequence

$$g(x) = \sum_{s=1}^q \beta_s x^s.$$

159

10.3. As the variates y, xy, x^2y, $\ldots$, x^qy in (10.9) can all be regarded as embodying in their observed values departures from a relationship, it would seem that an alternative way of regarding this model would be as a functional relationship model. Consider whether an approach along the lines of that developed in section 6.2 could be applied to this case.

10.4. For what values of θ does (10.17) give rise to (i) the exponential curve and (ii) the Gompertz curve?

$$\text{(Richards (1959), Nelder (1961)).}$$

References

ACTON, F. S. (1959). *Analysis of Straight Line Data*. Wiley, New York.

AITKEN, A. C. (1933). 'On fitting polynomials to data with weighted and correlated errors. *Proc. Roy. Soc. Edin.*, **54**, 12–16.

AITKEN, A. C. (1934). 'On least squares and linear combinations of observations'. *Proc. Roy. Soc. Edin.*, **55**, 42–48.

ANDERSON, T. W. (1958). *Introduction to Multivariate Statistical analysis*. Wiley, New York.

ANDERSON, T. W. (1963). 'Asymptotic theory for principal component analysis'. *Ann. Math. Statist.*, **34**, 122–148.

ANSCOMBE, F. J. (1967). 'Topics in the investigation of linear relations fitted by the method of least squares'. *J. Roy. Statist. Soc. B*, **29**, 1–52.

BARNETT, V. D. (1967). 'A note on linear functional relationships when both residual variances are known'. *Biometrika*, **54**, 670–672.

BARTLETT, M. S. (1937). 'Some examples of statistical methods of research in agriculture and applied biology'. *J. Roy. Statist. Soc.*, suppt. **4**, 137–183.

BARTLETT, M. S. (1947). 'Multivariate analysis'. *J. Roy. Statist. Soc.*, suppt., **9**, 176–197.

BARTLETT, M. S. (1949). 'Fitting a straight line when both variables are subject to error'. *Biometrics*, **5**, 207–212.

BEALE, E. M. L. (1960). 'Confidence regions in non-linear estimation'. *J. Roy. Statist. Soc. B*, **22**, 41–88.

BEALE, E. M. L. (1966). 'Discussion on Sprent (1966)'. *J. Roy. Statist. Soc. B.*, **28**, 293.

BEALE, E. M. L., KENDALL, M. G. and MANN, D. W. (1967). The discarding of variables in multivariate analysis'. *Biometrika*, **54**, 357–366.

BERKSON, J. (1950). 'Are there two regressions?' *J. Amer. Statist. Assoc.*, **45**, 164–180.

BERKSON, J. (1953). 'A statistically precise and relatively simple method of estimating the bio-assay with quantal response based upon the logistic function'. *J. Amer. Statist. Soc.*, **48**, 565–599.

BOWDEN, D. C. and GRAYBILL, F. A. (1966). 'Confidence bands of uniform and proportional width for linear models'. *J. Amer. Statist. Assoc.*, **61**, 182–198.

BOX, G. E. P. and LUCAS, H. L. (1959) 'Design of experiments in non-linear situations'. *Biometrika*, **46**, 77–90.

BROWN, R. L. (1957) 'Bivariate structural relation'. *Biometrika*, **44**, 84–96.

CARLSON, F. D., SOBEL, E., and WATSON, G. S. (1966). 'Linear relationships between variables affected by errors'. *Biometrics*, **22**, 252–267.

CHERNOFF, H. (1953). 'Locally optimal designs for estimating parameters'. *Ann. Math. Statist.*, **24**, 586–602

COCHRAN, W. G. (1938). 'The omission or addition of an independent variate in multiple linear regression.' *J. Roy. Statist. Soc., suppt.*, **5**, 171–176.

COCHRAN, W. G. (1957). 'Analysis of covariance—its nature and uses'. *Biometrics*, **13**, 261–281. (This issue of *Biometrics* was devoted to analysis of covariance.)

COCHRAN, W. G. and COX, G. M. (1957). *Experimental Designs*. 2nd. edn., Wiley, New York.

COX, D. R. (1968). 'Notes on some aspects of regression analysis'. *J. Roy. Statist. Soc. A*, **131**, 265–279.

COX, D. R. and SNELL, E. J. (1968). 'A general definition of residuals'. *J. Roy. Statist. Soc., B*, **30**, 248–279.

CREASY, M. A. (1956). 'Confidence limits for the gradient in the linear functional relationship'. *J. Roy. Statist. Soc., B*, **18**, 65–69.

DANIEL, C. (1959). 'Use of half normal plots in interpreting factorial two level experiments'. *Technometrics*, **1**, 311–341.

DAVIES, O. L. (1954). *The Design and Analysis of Industrial Experiments*. Oliver and Boyd, Edinburgh.

DAVIS, P. J. (1963). *Interpolation and Approximation*. Blaisdell, Mass.

DE LURY, D. B. (1950). *Values and Integrals of the Orthogonal Polynomials up to $n = 26$*. University of Toronto Press.

DRAPER, N. R. and SMITH, H. (1966). *Applied Regression Analysis*. Wiley, New York.

DUNN, O. J. (1968). 'A note on confidence bands for a regression line over a finite range'. *J. Amer. Statist. Assoc.*, **63**, 1028–1033.

EHRENBERG, A. S. C. (1968). 'The elements of lawlike relationships. *J. Roy. Statist. Soc. A*, **131**, 280–302.

ELSTON, R. C. and GRIZZLE, J. E. (1962). 'Estimation of time response curves and their confidence bands'. *Biometrics*, **18**, 148–159.

FIELLER, E. C. (1940). 'The biological standardization of insulin.' *J. Roy. Statist. Soc., suppt.*, **7**, 1–64.

FIELLER, E. C. (1944). 'A fundamental formula in the statistics of biological assay, and some applications'. *Quart. J. Pharm.*, **17**, 117–123.

FINNEY, D. J. (1952). *Probit. Analysis*. Cambridge University Press.

FISHER, R. A. and YATES, F. (1963). *Statistical Tables for Biological, Agricultural and Medical Research*. 6th edn., Oliver and Boyd, Edinburgh.

FISK, P. R. (1966). *Stochastically Dependent Equations*. Griffin, London.

FISK, P. R. (1967). 'Models of the second kind in regression analysis'. *J. Roy. Statist. Soc., B*, **29**, 266–281.

REFERENCES

GALTON, F. (1886). 'Family likeness in stature'. *Proc. Roy. Soc. Lond.*, **40**, 42–72.

GARSIDE, M. J. (1965). 'The best sub-set in multiple regression analysis'. *App. Statist.* **14**, 196–200.

GAUSS, C. F. (1809). *Theoria motus corporum coelestium in sectionibus conicis solem ambientum.* Pathes and Besser, Hamburg. (English translation by C. H. Davis, Boston, 1859).

GOMES, F. P. (1953). 'The use of Mitscherlich's regression law in the analysis of experiments with fertilizers'. *Biometrics*, **9**, 498–516.

HALPERIN, M. and GURIAN, J. (1968). 'Confidence bands in linear regression with constraints in the independent variables'. *J. Amer. Statist. Assoc*, **63**, 1020–1027.

HEALY, M. J. R. (1963). 'Fitting a quadratic'. *Biometrics*, **19**, 362–363.

HEALY, M. J. R. (1966). 'Discussion on Sprent (1966)'. *J. Roy. Statist. Soc.*, B, **28**, 290–291.

HEALY, M. J. R. (1968). 'Multiple regression with a singular matrix.' *Applied Statistics*, **17**, 110–117.

HEY, E. N. and HEY, M. H. (1960). 'The statistical estimation of a rectangular hyperbola'. *Biometrics*, **16**, 606–617.

HIORNS, R. W. (1965). 'The fitting of growth and allied curves of the asymptotic regression type by Stevens's method'. *Tracts for Computers, XXVIII*, University College, London. Cambridge University Press.

HOCKING, R. R. and LESLIE, R. N. (1967). 'Selection of the best sub-set in regression analysis'. *Technometrics*, **9**, 531–540.

HOEL, P. G. (1962). *Introduction to Mathematical Statistics.* 2nd edn. Wiley, New York.

HOPKINS, J. W. (1966). 'Some considerations in multivariate allometry'. *Biometrics*, **22**, 747–760.

HOTELLING, H. (1936). 'Relations between two sets of variates'. *Biometrika*, **28**, 321–377.

HUXLEY, J. S. (1924). 'Constant differential growth ratios and their significance'. *Nature, Lond.*, **114**, 895–896.

KEMPTHORNE, O. (1952). *The Design and Analysis of Experiments.* Wiley, New York.

KENDALL, M. G. (1951). 'Regression, structure and functional relationship I'. *Biometrika*, **38**, 11–25.

KENDALL, M. G. (1952). 'Regression, structure and functional relationship II'. *Biometrika*, **39**, 96–108.

KENDALL, M. G. (1957). *A Course in Multivariate Analysis.* Griffin, London.

KENDALL, M. G. and STUART, A. (1967). *The Advanced Theory of Statistics*, Vol. II., 2nd edn. Griffin, London.

KIEFER, J. and WOLFOWITZ, J. (1956). 'Consistency of the maximum likelihood estimators in the presence of infinitely many incidental parameters'. *Ann. Math. Statist.*, **27**, 887–906.

163

LEECH, F. B. and HEALY, M. J. R. (1959). 'The analysis of experiments on growth rate'. *Biometrics*, **15**, 98–106.

LINDLEY, D. V. (1947). 'Regression lines and linear functional relationship'. *J. Roy. Statist. Soc., suppt.*, **9**, 218–244.

LINDLEY, D. V. (1968). 'The choice of variables in multiple regression'. *J. Roy. Statist. Soc. B*, **30**, 31–66.

LINDLEY, D. V. and EL-SAYYAD, G. M. (1968). 'The Bayesian estimation of a linear functional relationship'. *J. Roy. Statist. Soc. B*, **30**, 190–202.

MADANSKY, A. (1959). 'The fitting of straight lines when both variables are subject to error'. *J. Amer. Statist. Assoc.*, **54**, 173–205.

MALINVAUD, E. (1966). *Statistical Methods of Econometrics*. (Translated by A. Silvey.) North Holland Publishing Company, Amsterdam.

MITCHELL, A. F. S. (1968) 'Exponential regression with correlated observations'. *Biometrika*, **55**, 149–162.

MOOD, A. M. and GRAYBILL, F. A. (1963). *Introduction to the Theory of Statistics*. 2nd edn., McGraw-Hill, New York.

NELDER, J. A. (1961). 'The fitting of a generalization of the logistic curve'. *Biometrics*, **17**, 89–110.

NELDER, J. A. (1962). 'An alternative form of a generalized logistic function.' *Biometrics*, **18**, 614–616.

NELDER, J. A. (1965). 'The analysis of randomized experiments with orthogonal block structure. I and II. *Proc. Roy. Soc. Lond. A*, **283**, 147–178.

NELDER, J. A. (1968). 'Regression, model building and invariance. *J. Roy. Statist. Soc. A.*, **131**, 303–329.

NELDER, J. A. and MEAD, R. (1965). 'A simplex method for functional minimization'. *Computer J.*, **7**, 308–313.

NEWTON, R. G. and SPURRELL, D. J. (1967). 'Examples of the use of elements for clarifying regression analysis.' *Applied Statistics*, **16**, 165–171.

NEYMAN, J. and SCOTT, E. L. (1951). 'On certain methods of estimating the linear structural relationship.' *Ann. Math. Statist.*, **22**, 352–361.

PATTERSON, H. D. (1956). 'A simple method for fitting an asymptotic regression curve'. *Biometrics*, **12**, 323–329.

PATTERSON, H. D. (1960). 'A further note on a simple method for fitting an exponential curve'. *Biometrika*, **47**, 177–180.

PATTERSON, H. D. (1969). 'Baule's equation'. *Biometrics*, **25**, 159–164.

PEARCE, S. C. (1963). 'The use and classification of non-orthogonal designs'. *J. Roy. Statist. Soc. A.*, **126**, 353–377.

PEARCE, S. C. (1965). *Biological Statistics, an Introduction*. McGraw-Hill, New York.

PEARSON, J. C. G. and SPRENT, P. (1968). 'Trends in hearing loss associated with age or exposure to noise.' *Applied Statistics*, **17**, 205–215.

PLACKETT, R. L. (1960). *Regression Analysis*. Clarendon Press, Oxford.

REFERENCES

POTTHOFF, R. F. and ROY, S. N. (1964). 'A generalized analysis of variance model useful especially for growth curve problems'. *Biometrika*, **51**, 313–326.

QUENOUILLE, M. H. (1959). 'Tables of random observations from standard distributions'. *Biometrika*, **46**, 178–202.

RAO, C. R. (1959). 'Some problems involving linear hypotheses in multivariate analysis'. *Biometrika*, **46**, 49–58.

RAO, C. R. (1965a). 'The theory of least squares when the parameters are stochastic and its application to the analysis of growth curves'. *Biometrika*, **52**, 447–458.

RAO, C. R. (1965b). *Linear Statistical Inference and Its Applications.* Wiley, New York.

RICE, J. R. (1964). *The Approximation of Functions.* Vol. 1. Addison-Wesley, Reading, Mass.

RICHARDS, F. J. (1959). 'A flexible growth function for empirical use'. *J. Expt. Bot.*, **10**, 290–300.

SEAL, H. (1964). *Multivariate Statistical Analysis for Biologists.* Methuen, London.

SEAL, H. (1967). 'Studies in the history of probability and statistics, XV. The historical development of the Gauss linear model'. *Biometrika*, **54**, 1–24.

SEARLE, S. R. (1966). *Matrix Algebra for the Biological Sciences.* Wiley, New York.

SPRENT, P. (1961). 'Some hypotheses concerning two phase regression lines'. *Biometrics*, **17**, 634–645.

SPRENT, P. (1965). 'Fitting a polynomial to correlated, equally spaced observations.' *Biometrika*, **52**, 275–276.

SPRENT, P. (1966). 'A generalized least-squares approach to linear functional relationships. *J. Roy. Statist. Soc. B*, **28**, 278–297.

SPRENT, P. (1967). 'Estimation of mean growth curves for groups of organisms'. *J. Theoret. Biol.*, **17**, 159–173.

SPRENT, P. (1968). 'Linear relationships in growth and size studies'. *Biometrics*, **24**, 639–656.

STEVENS, W. L. (1951). 'Asymptotic regression'. *Biometrics*, **7**, 247–267.

SWINDEL, B. F. (1968). 'On the bias of some least-squares estimators of variance in a general linear model'. *Biometrika*, **55**, 313–316.

TOCHER, K. D. (1952). 'The design and analysis of block experiments'. *J. Roy. Statist. Soc. B*, **14**, 45–100.

TUKEY, J. W. (1951). 'Components in regression'. *Biometrics*, **7**, 33–69.

TURNER, M. E., MONROE, E. J., and LUCAS, H. L. (1961). 'Generalized asymptotic regression and non-linear path analysis.' *Biometrics*, **17**, 120–143.

VILLEGAS, C. (1961). 'Maximum likelihood estimation of a linear functional relationship'. *Ann. Math. Statist.*, **32**, 1048–1062.

WALD, A. (1940). 'The fitting of straight lines if both variables are subject to error'. *Ann. Math. Statist.* **11**, 284–300.

WATSON, G. S. (1955). 'Serial correlation in regression analysis. I.' *Biometrika*, **42**, 327–341.

WATSON, G. S. (1967). 'Linear least squares regression. *Ann. Math. Statist.*, **38**, 1679–1699.

WEATHERBURN, C. E. (1946). *A First Course in Mathematical Statistics.* Cambridge University Press.

WETHERILL, G. B. (1967). *Elementary Statistical Methods.* Methuen, London.

WILLIAMS, E. J. (1955). 'Significance tests for discriminant functions and linear functional relationships'. *Biometrika*, **42**, 360–381.

WILLIAMS, E. J. (1959). *Regression Analysis.* Wiley, New York.

WILLIAMS, E. J. (1967). 'The analysis of association among many variates'. *J. Roy. Statist. Soc. B*, **29**, 199–242.

WISHART, J. (1938). 'Growth rate determinations in nutrition studies with the bacon pig and their analysis'. *Biometrika*, **30**, 16–28.

WISHART, J. and METAKIDES, T. (1953). 'Orthogonal polynomial fitting'. *Biometrika*, **40**, 361–369.

Author Index

Subject Index